ANALYSIS
What Analytical Chemists Do

Royal Society of Chemistry Paperbacks

Royal Society of Chemistry Paperbacks are a series of inexpensive texts suitable for teachers and students and giving a clear, readable introduction to selected topics in chemistry. They should also appeal to the general chemist. For further information on selected titles contact:

Sales and Promotion Department
The Royal Society of Chemistry
Burlington House
Piccadilly
London W1V 0BN

Telephone: 01–734 9864

Titles Available

Water *by Felix Franks*
Food – The Chemistry of Its Components *by T.P. Coultate*
Analysis – What Analytical Chemists Do *by Julian Tyson*

How to Obtain RSC Paperbacks

Existing titles may be obtained from the address below. Future titles may be obtained immediately on publication by placing a standing order for RSC Paperbacks. All orders should be addressed to:

The Royal Society of Chemistry
Distribution Centre
Blackhorse Road
Letchworth
Herts. SG6 1HN

Telephone: Letchworth (0462) 672555
Telex: 825372

Royal Society of Chemistry Paperbacks

ANALYSIS
What Analytical Chemists Do

JULIAN TYSON

Department of Chemistry
University of Technology, Loughborough

ROYAL
SOCIETY OF
CHEMISTRY

British Library Cataloguing in Publication Data

Tyson, Julian
    Analysis : what analytical chemists do. –
    (Royal Society of Chemistry Paperbacks,
    ISSN 0262–9518).
    1. Chemistry, Analytic
    I. Title
    543   QD75.2

ISBN 0–85186–463–5

Published by The Royal Society of Chemistry
Burlington House, Piccadilly, London W1V 0BN

Typeset by David John Services Ltd., Slough
Printed by Adlard and Son Ltd., Letchworth

Preface

This book is for students, but it is not meant to be a textbook to be used in conjunction with a formal teaching programme. Some readers may feel therefore that the material is not presented in the most logical fashion. The order of the material reflects my attempts, firstly, to describe what chemists and analytical chemists do and, secondly, to explain how analytical chemists do it.

Devising methods to obtain information about the chemical composition of materials requires an understanding of some chemistry as well as of some of the physics of measurement, and so much of the book is given over to explanations of the relevant aspects of these two sciences. I have endeavoured to explain the topics at a level appropriate to the readership, which means I am treading the thin line between over-simplification and unnecessary complexity. I hope that readers will bear with me when they feel I have strayed from this line (in whatever direction). Some topics are explained in more detail than others. This reflects the extent to which I think these topics are used by analytical chemists. Many of these more detailed explanations are contained in 'Boxes' and may therefore be readily identified and omitted – at the first reading only, of course.

A number of people have helped in the production of this book and I gratefully acknowledge their contributions. Dr. R.A. Chalmers first opened my eyes to the fascination of the analytical aspects of chemistry during my period as an undergraduate at Aberdeen University, and Professor T.S. West made sure they were kept open during my period as a research student at Imperial College. My analytical colleagues at Loughborough

University provide an environment in which I have hardly blinked for the last ten years, and Professor J.N. Miller and Drs. A.G. Fogg, R.M. Smith, and T.E. Edmonds have been good enough to read and criticise the manuscript. In this latter respect I am also grateful to my colleague Dr. A.N. Strachan and to Mr. M.W. Riggall of Hind Leys Community College, Shepshed.

I am particularly grateful to my long-suffering family, Aileen, John, and Jennifer, who have allocated me much of the time necessary for this project.

Julian Tyson

Contents

Chapter 5
Tools of the Trade   105

Chapter 6
Problems with Mixtures –
Chemistry to the Rescue   136

Chapter 7
Tackling the Problems

Boxes

Chapter 1

What Do Analytical Chemists Do?

Chemicals are part and parcel of our everyday life. The quality of life we enjoy in our part of the world in the latter part of the twentieth century heavily depends on the chemical industry. The basic materials for fabrication – metals, cements, glass, and plastics – are all produced by the thousands of tonnes each year. Even wooden objects are glued, nailed, or screwed together by a product of a sector of the chemical industry, and are maybe finished off with a thin layer of varnish or paint.

Clothes consist, in part at least, of a manufactured fibre and are coloured by dyes and washed in detergents that have been formulated and made by chemists. As well as keeping us clean and warm and dry, chemicals are closely involved in keeping us alive by virtue of what we eat and drink. Agriculture could not have achieved its present level of output without fertilisers, pesticides, animal feeds, supplements, *etc.*, nor could ill-health be combated so effectively without the enormous range of medicines produced by the pharmaceutical industry.

If you make a list of all the things you can see now that are manufactured chemicals and compare it with the corresponding list of natural materials, you will perhaps appreciate just how important the modern chemical industry is. It is strange that the public at large has come to regard the word 'chemical' as meaning something nasty and dangerous, whereas the reality of the matter is that we are now almost totally dependent on the skill and ingenuity of our chemists.

An even more sobering thought is that we are totally dependent on a continuing supply of the starting materials. Skilled

1

though chemists are, even they are subject to the law of conservation of matter. Ballpoint pens and semi-conductor devices are not made out of thin air, nor do the starting materials grow on trees. Though who knows: when what we can get out of the ground runs out, cellulose may come into its own as a feedstock material since the most efficient way of chemically storing solar energy is at present by green-plant photosynthesis. At the moment it is oil and mineral deposits that keep the chemical industries going. But not just the chemical industry, these materials also provide our energy. In the not too distant future we will be faced with the problem of what to do when these resources run out. Although it is difficult to predict accurately when these basic feedstocks will be depleted, it will be up to chemists to provide solutions to the problems this will cause.

At present there are about 50 000 chemists in the U.K. (approximately half of all the U.K. scientists). The major employers of chemists are shown in Figure 1.1, the various major categories of jobs that chemists do are shown in Figure 1.2, and a further, more detailed breakdown of the variety of manufacturing industries, government, and other service industries in which chemists are employed is shown in Figure 1.3. No doubt you

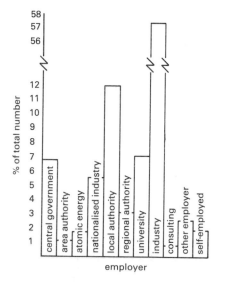

Figure 1.1 *Relative numbers of chemists working for their major employers in the U.K.*

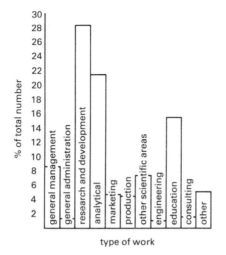

Figure 1.2 *The relative numbers of chemists in the major types of work that they do. Note that analytical work is one of the three major categories of work for chemists.*

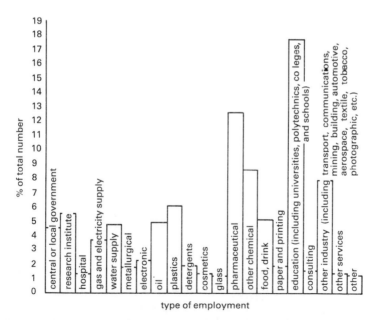

Figure 1.3 *A more detailed breakdown of the field of employment of chemists in the U.K.*

appreciate the role of the chemist in agriculture, cement manufacture, cosmetic formulation, food preparation, packaging technology, petroleum refining, drug synthesis, *etc.* But what about
analytical chemists – where do they fit into the picture?

THE IMPORTANCE OF CHEMICAL COMPOSITION

To appreciate the role of the analytical chemist, we need to look a
bit more closely at the materials produced by the chemical
industry. All industrial products, whether they are to be viewed as
chemicals or as something more complex, are designed for a
particular purpose or to do a specific job. It may be quite a simple
job such as providing a relatively rigid, insoluble container for
hot, aqueous liquids that won't break when dropped on the
kitchen floor (a plastic cup), or it may be a more complicated job
such as the combination of the long list of specifications for the
components of a video recorder. The fundamental design
feature common to both these examples is that the components
and the complete article will only be up to specification if they
have the correct chemical composition. That is to say the
mechanical, electrical, optical, thermal, and other properties of a
material depend on its chemical nature. In some cases it is not just
the relative proportions of the various atoms or molecules that
are important but also the way in which the atoms are linked
together. This is particularly true in the case of polymers whose
mechanical properties (flexibility, rigidity, *etc.*) are directly related to the extent of crosslinking between polymeric chains. In
other cases the properties may depend on relatively small
proportions of constituent atoms or molecules. In steel the minor
alloying elements, carbon, tungsten, nickel, cobalt, *etc.*, govern
the mechanical properties and even, in the case of chromium, the
chemical properties (*e.g.* corrosion resistance). A more striking
example of the role of minor components is provided by the
electrical properties of semi-conductor devices, governed by key
elements such as arsenic or germanium, which are present at very
low concentrations indeed in the silicon.

The material may be a mixture of chemicals that only has the
right properties because of the composition of the mixture. A
processed food may contain preservatives, colouring, sugar,
vitamins, and emulsifiers added to the basic fat, protein, and/or
carbohydrate. Beer only has its intoxicating effect because of the

alcohol content, and its taste is due to the presence of chemicals extracted from the malt or hops during the brewing process.

Manufacturers obviously must not only be able to perform the appropriate chemical reactions to convert the starting materials into the desired product but also ensure that the product has the appropriate chemical composition. There may be a range of compositions for which the product has the required properties, and it is clearly economic sense to minimise the proportion of the expensive chemicals. As we shall see later, analytical chemists, too, are subject to financial constraints in what they do.

WHAT DO ANALYTICAL CHEMISTS DO?

What analytical chemists do, as you may already have guessed, is check the chemical composition of the products. However, this is not the only thing that analytical chemists do: there are many other things, apart from the products of various manufacturing processes, that need to have their chemical compositions checked. But let's start by considering this aspect of the analytical chemist's work.

As a very large number of materials are mixtures of chemicals, every manufacturing company needs to employ analytical chemists whose job it is to devise ways of measuring the relative amounts of the various chemical species that go to make up the particular material. This may be the concentration of elements like carbon and nickel in the case of a steel sample or the concentration of certain molecules such as the vitamins in a breakfast cereal or the amount of drug in a pill. Not only do analytical chemists have to devise ways of carrying out these measurements but they also have to be able to identify chemicals. This is because no manufacturing process runs indefinitely without a hitch and when something goes wrong analytical chemists will be called on to help troubleshoot. This may involve identifying the impurities or components that have thrown the process off beam and trying to track down their source. It may be that the nature of a starting material has changed, perhaps because the company has changed supplier or critical reaction conditions have not been maintained, or a catalyst has reached the end of its useful life.

If the company is selling the product in a very competitive market, the analytical chemists will be given the job of finding out the nature and composition of the competitor's products. Even when production is running smoothly, there may be the composition of various intermediates to check and the composition of the starting materials will have to be verified. The research labs will be making new materials that will have to be analysed, and the levels of toxic materials in effluents that a factory discharges into the atmosphere or into a water course will have to be monitored. The role of the analytical chemist in industry is summarised in Figure 1.4.

Figure 1.4 *The analytical chemist in industry:* (a) *sampling and testing raw materials,* (b) *testing intermediates,* (c) *monitoring product quality, and* (d) *monitoring effluent quality.*

What Else Do Analytical Chemists Do?

Apart from the role of the analytical chemist in the manufacturing and processing industries, there are many other aspects of our complex life-style in which analytical chemists (and other chemists, of course) are involved. For example, did you know that analytical chemists are ...

Helping to Save Lives. The laboratories of nearly all hospitals contain a considerable number of analytical personnel (called clinical chemists) who carry out a wide range of jobs. These involve the analysis of samples of patients' blood, urine, *etc.* for a variety of components to help doctors with their diagnoses (see Figure 1.5). The progress of a patient with a particular disease

Figure 1.5   *The analytical chemist in hospital.*

may be followed by monitoring the concentrations of certain key components of the blood or urine. The effectiveness of a course of drugs may be assessed by analysing for the particular active ingredient or its metabolites (compounds into which the body converts the drug in order to try to remove it). Often a drug may not be effective at very low levels but is harmful at high concentrations and so the patient must be checked regularly to see that the level of drug circulating in the body is correct. If you think of the number of people who have a 'sample' taken by their doctor or health clinic staff and of the number of people in hospital or being treated with drugs at any one time, you begin to get an idea of the problems that clinical chemists face. Dealing with a very large number of samples and producing quick answers (to the questions the doctors are asking in this case) is a problem that will be discussed later. In cases of poisoning, a specialist analytical chemist called a toxicologist will be involved who may have to identify the toxic material responsible but may have little evidence to work with if the victim is unconscious.

Helping to Protect the Consumer and the Environment. The quality of some of the materials that we make use of in our everyday lives (*e.g.* those we eat and drink) is controlled by acts of parliament. Among other jobs, the public analyst checks all these products regularly to see if they conform to the appropriate legal requirement (see Figure 1.6). The work of the public analysts and their

Figure 1.6 *The analytical chemist checks* (a) *food and drink and* (b) *the environment.*

staff is obviously very important and the highest standards are required, to the extent that public analysts must have a special qualification in analytical chemistry (called the Mastership of Chemical Analysis). Central and local government have large numbers of scientists working for them, including a good proportion of analytical chemists. The best-known establishment is probably the Laboratory of the Government Chemist (LGC) in London. The scientific civil servants who work there not only analyse food, drink, medicines, pesticides, *etc.* but also provide a service to Customs and Excise by examining beers, wines, spirits, oils, petrol, tobacco products, and a variety of contraband including drugs.

Along with the Health and Safety Executive laboratories, the LGC also monitors the level of toxic and hazardous substances in the work-place environment, such as the level of asbestos fibres, mercury vapour, or lead dust. Like the clinical chemist, the scientific civil servant in this situation may be faced with the problem of measuring very small amounts of the potentially hazardous chemical as well as dealing with a work-load of an extremely large number of analyses.

The provision of drinking water is a process which must be carefully and constantly monitored. The Water Authorities make great use of analytical chemists' skills in keeping a check on the quality of the water, not only in terms of the materials that must be removed before the water is considered drinkable but also in terms of the chemicals that are added, such as fluoride to help keep our teeth in good shape and chlorine to kill bacteria. The water analyst will also be keeping an eye on the contents of surface waters and on what is discharged into the sewerage system, particularly those into which industrial plants discharge effluent (see Figure 1.6). The sudden arrival of a high concentration of a transition metal at the sewage works, for example, can kill the bacteria used in the purification process. In the normal course of events, the sludge produced in the process is often used as a fertiliser, so it is important that the concentration of potentially toxic metals is known, as these can have harmful effects on crops. As with the action of drugs mentioned earlier, there may well be a concentration range in which the action of the metal is beneficial, even desirable. The organism may suffer ill-health at metal concentrations both below this range, because of a deficiency, and above it, because of a toxic effect.

Helping the Farmer.   The role of agricultural chemists in increasing the efficiency of farming has already been mentioned. Analytical chemists play a role in monitoring the levels of nutrients in soils, giving the farmer information of use in deciding how much of which fertiliser to apply (see Figure 1.7). There are a number of laboratorie around the country providing such a service to farmers through the skills of soil scientists and agricultural analytical chemists. Levels of potentially harmful materials will be monitored, not only in the soil but also in crops and animals. It is also important to monitor levels of pesticides, weed killers, *etc.* applied to crops and soils that could pass along the food chain to us. The Ministry of Agriculture, Fisheries, and

Figure 1.7 *The analytical chemist helps the farmer.*

Foods has laboratories which monitor the residual concentrations of some of the components of animal feeding stuffs or veterinary drug residues in meat. Detection and measurement of some of these components are often extremely difficult as they may be present at very low concentrations, and the requirements of the analysis place severe demands on the analytical chemist's skill and ingenuity.

Helping to Catch Criminals. The work of the police in solving crimes and bringing the culprits to justice is considerably helped by analytical chemists. The specialised analysts involved in this type of work are known as forensic (meaning 'related to the courts of law') scientists. A lot of their work is concerned with estab-

Figure 1.8 *The analytical chemist helps to catch criminals.*

lishing a contact, *i.e.* was such-and-such a person at such-and-such a place or was this vehicle at this place and so on (see Figure 1.8). Answering this type of question is based on the principle that 'every contact leaves its traces', and the work may involve microscopic examination and the measurement of several of the chemical components. For example, when a hand gun is fired, the hand of the firer receives a small amount of powder, ejected from the muzzle and breech, arising from the propellant in the bullet. This material has a high concentration of antimony and barium, so if the hands of a suspect are swabbed and an unusually high concentration of these elements is found then it *may* mean that the person has recently fired a gun.

Victims of hit-and-run incidents may have fragments of paint or glass lodged in their clothing. Comparison of the trace-element concentrations in a glass fragment taken from the victim's clothing with the trace-element concentrations in glass from the headlamp of a suspect vehicle may establish that the two materials *could* be the same. Different types of glass have different trace-element concentrations arising from variations in starting materials or in the manner of manufacture; so for example window glass is easily distinguished from headlamp glass. How easily one headlamp is distinguished from another is a problem with which the forensic scientist has to grapple in putting together a case to take to court. There are lots of other materials that may be transferred during a contact, such as clothing fibres, dust, soil, blood, and so on. Very often the forensic scientist is faced with the problem that the amount of material available for analysis is very small so that even if the components which characterise the material are major components the final analysis may involve very low concentrations. For example, in the analysis of gunshot residues the amount of the material deposited is so small that even if the final volume of swab liquid is 0.5 cm^3 the concentration of the metals may only be at the part per million (p.p.m.) level. A 'p.p.m.' is a concentration equivalent to 1 µg (*i.e.* 10^{-6} g) in 1 cm^3. The limited amount of material means that there is no room for mistakes: the forensic scientist cannot ask for more if the solution is spilled or contaminated accidentally, as you can in the teaching laboratory.

Helping to Ensure Fair Play.   It is a sad reflection on the position that sport occupies in our society that some competitors resort to artificial means to increase their performance (see Figure 1.9). In

Figure 1.9 *The analytical chemist helps to detect drug abuse in sport.*

some cases the competitor may have no say in the matter, as for example in the case of a racehorse or greyhound. The abuse of drugs in connection with sporting events is sufficiently wide-spread for a number of laboratories, admittedly a small number, to be involved in checking blood and urine samples for drugs of this type or their metabolites. The problems are formidable, as the material being sought may be present only at very low concentrations in a material which is itself a complex mixture of chemical substances. How the analytical chemist deals with such complex mixtures will be explained in a later chapter.

Helping in Other Ways. The list of the jobs that analytical chemists get up to is not intended to be exhaustive, and you can probably think of other aspects of our lives that analytical chemists will be involved in, such as geological surveys, prospecting, coal and gas supply, electricity generation, and so on. There are even a few of us involved in the teaching and training of analytical chemists.

THE GENERAL PHILOSOPHY OF ANALYTICAL CHEMISTRY

Analytical chemists provide information, on which to base a decision, about the chemical composition of bulk materials. The various processes involved in producing such information are outlined in Figure 1.10. It is the analytical chemist's job to select the most appropriate links in the chain from 'bulk material' to 'decision', and then to ensure that the analysis is carried out

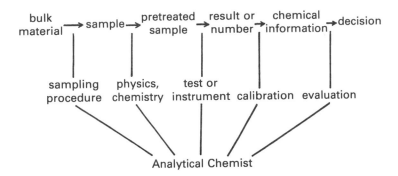

Figure 1.10 *The various stages in the overall analytical method.*

competently. The methods and techniques available to the analytical chemist and the various constraints which need to be considered in devising an overall analytical procedure are discussed in the later chapters. Devising such analytical procedures is not a trivial matter and, in addition to problems requiring chemical skill and ingenuity, there are other problems to be considered, such as how to deal with a very large number of samples or with very small samples and how likely is the result of the analysis to be correct. The concentration of the chemical species being sought may be vanishingly small or enormously large, and it is unlikely that the same analytical method will apply to both situations. In addition, the cost of the analysis and the speed with which it can be done may be factors of prime importance.

Most of the rest of this book is devoted to explaining how aspects of chemistry may be used to provide answers to the questions that analytical chemists are asked about the composition of bulk materials.

Although the professional analytical chemist will be responsible for the quality of the information produced and thus must have control over every stage in the procedure summarised in Figure 1.10, in practice the initial sampling may be done by a person other than the analytical chemist. This is because the bulk material may be located some distance from the analytical laboratory and its composition may be changing with time. The next chapters consider what goes on in the analytical laboratory, *i.e.* what happens to the sample once it has been taken and delivered to the lab.

The analytical chemist will not forget that if this sample is not representative of the bulk material then any decision based on the information provided as a result of the laboratory procedures may be unsound.

FURTHER READING

B. Dixon, 'What Is Science For?', Penguin, Harmondsworth, 1976.

A. Isaacs, 'Introducing Science', Penguin, Harmondsworth, 1972.

'Chemistry and the Needs of Society', Special Publication No. 26, The Chemical Society, London, 1974.

M.D. Wynne, 'Chemical Processing in Industry', Monograph for Teachers No. 16, The Royal Institute of Chemistry, London, 1970.

D.F. Ball, 'Some Aspects of Technological Economics', Monograph for Teachers No. 25, The Chemical Society, London, 1974.

'Chemistry and Agriculture', Special Publication No. 36, The Chemical Society, London, 1979.

'Understanding Our Environment', ed. R.E. Hester, The Royal Society of Chemistry, London, 1986.

'Pollution: Causes, Effects and Control', ed. R.M. Harrison, Special Publication No. 44, The Royal Society of Chemistry, London, 1983.

G.F. Lewis, 'Analytical Chemistry', 2nd Edn., Macmillan, London, 1985.

R.A. Chalmers, 'Aspects of Analytical Chemistry', Oliver and Boyd, Edinburgh, 1968.

D. Betteridge and H.E. Hallam, 'Modern Analytical Methods', The Chemical Society, London, 1972.

G.E. Baiulescu, C. Patroescu, and R.A. Chalmers, 'Education and Teaching in Analytical Chemistry', Ellis Horwood, Chichester, 1982.

Chapter 2
Making Light Work

Examples of some of the materials about which analytical chemists are requested to provide information are given in the previous chapter. Thus analytical chemists are asked to provide information not only about the nature of the components of a mixture, *i.e.* qualitative analysis (the identification of the chemicals), but also about how much of a particular component is present, *i.e.* quantitative analysis.

Both categories of information may be provided by the application of the way that light and the material in question interact. For many of us the attraction of chemistry is our perception of chemical reactions through the senses of sight, smell, and hearing. In particular, the colour effects produced by many chemical reactions are a continued source of fascination. There are many analytical situations where the perception of the appropriate chemistry directly by eye is adequate to provide the required information. There are fewer examples in which noses and ears will do the same. However, there are also a great many analytical problems, for which the interaction of light is the basis of the appropriate solution, where the perceptive skill of the eye is inadequate. In these situations analytical chemists make use of an appropriate instrument that in effect extends the abilities of the human eye. One such situation is the requirement to quantify low light intensities or small changes in light intensity. This in turn has arisen from the requirement to provide information about components of a mixture present at very low concentrations. The analysis of such components is known as trace analysis.

TRACE ANALYSIS

In general terms, 'trace' components are present at concentrations ranging from 0.01% to 0.000 000 01% on a mass basis in a solid material. Percentages are cumbersome to deal with at these low concentrations because of all the zeros between the decimal point and the first significant figure, and it is more convenient to work with p.p.m., parts per million (*i.e.* $\mu g\ g^{-1}$ or mg kg^{-1}), or p.p.b., parts per American billion (*i.e.* ng g^{-1} or $\mu g\ kg^{-1}$). Thus the range above would be 100 mg kg^{-1} to 0.1 $\mu g\ kg^{-1}$. Below this level we would speak of the microtrace level (100 ng kg^{-1} to 100 pg kg^{-1}). Some analyses are possible at even lower concentrations than this, the nanotrace level, but for the time being we will give them a miss. The p.p.b. level is formidable enough when you realise that a family of four people constitutes about 1 p.p.b. of the entire population of the planet. In some respects the analytical chemist's job is equivalent to that of an alien visitor who is trying to find you.

The number of and requirement for trace analyses are continually growing and the problems of trace analysis are probably the biggest challenge facing analytical chemists at present. Demands for the identification and measurement of an ever increasing number of chemical species at ever decreasing concentration levels are coming from an ever increasing number of sources. Considerable effort is being devoted to devising analytical chemistry techniques that can solve some of these problems. Much of this effort concerns the application and development of techniques based on the interaction of electromagnetic radiation with matter (known as spectroscopic techniques or, simply, spectroscopy), and much of this chapter is thus devoted to such techniques.

However, we will start with familiar ground in order to illustrate some basic concepts of qualitative and quantitative analysis. And we need to have some idea as to how light interacts with chemicals in order to understand how the analytical chemist can exploit the various phenomena to provide analytical information. So, although the quantitative analytical techniques you may be most familiar with are titrimetry and gravimetry, we will leave these to a later chapter because spectroscopic techniques occupy key roles in most analytical laboratories. There are a large number of different kinds of analytical spectroscopy. These range from the use of γ-rays (frequency 10^{22} Hz, wavelength 3 ×

10^{-14} m) to the use of radio waves (frequency 10^5 Hz, wavelength 3×10^3 m), and instruments using these and the various parts of the spectrum in between are in routine use in analytical laboratories. However, before we look at some of these instruments and the information they provide, let us start with a familiar part of the spectrum, namely the range of wavelengths between about 4.00×10^{-7} m and 7.50×10^{-7} m (400 to 750 nm), which is referred to as the visible spectrum or visible light.

LIGHT FANTASTIC

The interaction of light with materials is such a common event for us that we barely give it a second thought. But, in fact, it is an important factor in establishing the quality of our lives. The most common form of interaction that we 'see' is the absorption of light. This is the process which gives a material its colour. The light reflected (*i.e.* that which is not absorbed) from the surface of an object differs in its composition of wavelengths from that received directly from the light source (the sun, tungsten filament bulb, fluorescent tube, *etc.*), and our eye passes 'messages' to our brain that we interpret as colour. Colour plays an important role in our lives: we go to extraordinary lengths to surround ourselves with colours and combinations of colours which we find pleasing and interesting. Production of such colours in all their various forms is a multi-million pound industry. We paint our houses, decorate our rooms and offices, carefully choose our clothes, paint our faces, dye our hair, fuss over the appearance of our new car, trade in our monochrome TV sets, plant flowers in our gardens; we celebrate with colour and mourn with the lack of it. Punishment environments are 'monochromous'; discotheques are the opposite.

Light Absorption and Qualitative Analysis

We commonly use the absorption of light by a material being examined to assess its quality. Consider the activity of the analyst shown in Figure 2.1. The sample under study is a liquid and it is being examined by transmitted light. The wavelengths of light transmitted compared with those which travel directly to the observer's eye enable some qualitative analysis to be performed. Firstly, from memory our analyst is able to recall what the pint of

Figure 2.1 *Examination of a material by transmitted radiation.*

beer looks like and to distinguish it from cider or lager. It may also be possible to confirm the identity of the particular type of beer. However, if there is an uncertainty about the result of this final step, the material under test may be compared with a reference material. In this case the companion's pint (deemed to be satisfactory by the companion – and supposedly identical to the material being examined) is used. This may prove a conclusive test, one way or the other. If there is still an uncertainty about the result, another analytical method may be used, rarely employed in laboratories these days but highly appropriate in this case, namely taste.

Qualitative analysis using spectroscopic techniques follows exactly the same process. The extent of the interaction of the sample with the electromagnetic radiation (in this case visible light) is measured (in this case by the human eye) and is compared with an authentic reference material (whose interaction is either measured or recalled). Alternatively, the extent of the interaction may be analysed in terms of the chemistry of the sample, provided one knows something about the way in which that particular type of radiation interacts with molecules, ions, *etc.*

Does the Sample Contain Lead? You may be already aware of schemes of qualitative analysis based on the separation of ions in solution into groups by particular reagents. In one such scheme the analysis for metals starts with the addition of hydrochloric acid to the sample solution to precipitate the insoluble chlorides of silver, lead(II), and mercury(I). Any precipitate obtained is then examined to see which metals are present. If lead is present, the precipitate, of $PbCl_2$, is redissolved in hot water and the hot solution is tested by adding some potassium iodide solution. If a yellow precipitate crystallising in silky plates is obtained, the presence of lead is indicated. The analytical chemistry test depends on the interaction of light with the chemical system, as we recognise the precipitate as lead iodide from its characteristic appearance. The certainty of the identification is considerably improved because the chemistry which has been performed on the sample has been designed specifically to isolate lead from other metals in the solution. Not only that, but the reaction conditions were chosen so that when the potassium iodide solution was added lead iodide was precipitated. There are no reagents which will always undergo a characteristic reaction with only the species being sought, giving a characteristic product, whether this is a precipitate, colour change, or whatever. Several generations of chemists have tried to produce such tests, and although there are reagents which are very nearly specific to one species (see Chapter 6) the ultimate aim has not been achieved.

As instrumental techniques have developed, research into this area of reagent chemistry has declined because, as we shall see later, the use of appropriate instruments can provide conclusive identification. There is nevertheless still a need for the non-instrumental chemical test as there are plenty of situations where the use of an instrument is inappropriate.

Let us imagine a water chemist investigating the possible contamination of a river by an accidental discharge of a lead solution. The appropriate instrument may not be in the back of the van, and there is not enough time to take a sample back to the lab and wait for the result. Adding potassium iodide solution to the sample of filtered river water is not a particularly reliable test as (*a*) there are other metals which give a precipitate with iodide [*e.g.* copper(I), mercury (I), and silver] and (*b*) lead iodide is soluble to the extent of 0.044 g per 100 cm^3 of cold water. This means that unless the concentration of lead is greater than about 0.2 g dm^{-3} (200 mg dm^{-3}) no precipitate will be formed. Adding a large excess of the reagent does not improve matters in this case as lead iodide is soluble in excess iodide, forming tetraiodoplumbate(II) ions, PbI_4^{2-}. A better test would be to take one drop of sample and to add a few drops of potassium cyanide solution (carefully, of course, as this solution is poisonous), a few drops of a solution of the sodium salt of 2-hydroxypropane-1,2,3-tricarboxylic acid (sodium citrate), and a drop of concentrated ammonia solution. Then shake with tetrachloromethane (carbon tetrachloride) saturated with diphenylthiocarbazone (dithizone). A red colour in the carbon tetrachloride layer indicates the presence of lead. Just under 1 mg dm^{-3} of lead would be detected.

The chemistry behind this is as follows. Dithizone $(C_6H_5-N=N-CS-NH-C_6H_5)$ forms very stable, very highly coloured compounds with a number of metals in alkaline solution, including lead. Neither dithizone nor the metal dithizonates are soluble in water and, so that the colour produced may be seen, an organic solvent is used. Cyanide, CN^-, forms more stable complexes with all of the metals likely to be present that would also give a red colour except iron(II), and it thus prevents their formation when the dithizone is added. Citrate forms a stable complex with iron(II) and so prevents it from reacting.

The test is not completely foolproof since the original test solution may have contained a chemical species which formed a more stable complex with the lead than does dithizone. This would give rise to a false negative result. An example of such an entity is 1,2-bis[bis(carboxymethyl)amino]ethane (ethylenediaminetetra-acetic acid or edta). Whether edta is likely to be present is something our water chemist friend will have to assess from a knowledge of what is being discharged into this section of the river.

All of this has taken us away somewhat from the starting point of the examination of beer by transmitted light, but it has done two things. It has provided an example of a trace analysis, namely the detection and identification of lead at concentrations down to 1 mg dm^{-3}. It has also illustrated the way in which the analytical chemist must use chemical knowledge in order (a) to devise a suitable qualitative analysis test and (b) to understand and appreciate the limitations to the test. This second part is very important because, for most chemical species, many tests have been devised and the analytical chemist is usually faced with the problem of which one to choose, rather than of designing the test procedure itself. In choosing the procedure several factors have to be taken into account (these will be discussed in Chapter 7), but interference effects, both positive [such as the effect of Cu(I) on the iodide test] and negative (such as the effect of edta), are obviously of fundamental importance.

How Much Lead Is Present?

We have already seen one possible approach to providing an answer to this question. If it is assumed that there are no other components of the sample that will interfere with the test, the absence of a lead iodide precipitate after the appropriate chemistry has been carried out must be interpreted as a lead concentration of below 200 mg dm^{-3}. Similarly, if no colour is formed with the dithizone test, then the concentration must be less than 1 mg dm^{-3}. The provision of answers like this may be sufficient in some circumstances, for example the testing of a solution of a chemical against a specification which allows for no more than 1 mg dm^{-3} lead. Such situations are often referred to as 'go–no-go' situations and the analytical methods as providing 'yes–no' answers. However, for a positive dithizone test it is possible to get much nearer the true concentration value than just 'greater than 1 mg dm^{-3}'. The way of doing this is based on the intensity of the colour in the solution.

The more concentrated a solution is, the darker the colour, *i.e.* the more light is absorbed. So, if a series of solutions whose concentrations are known is prepared and the colour-forming chemical reaction is carried out, a series of solutions with differing intensities of colours will be produced. These solutions are known as standards. The colour of the unknown sample solution is compared with the range of standard colours. The complete colorimetric procedure, as it is known, is illustrated in Figure 2.2.

The concentration of the unknown may then be specified as being a particular value (if the colour cannot be distinguished from one of the standards) or as being between two values (if the sample colour appears to be between the colours of two standards). It is important that the sample and standards are examined in the same-sized test-tubes or other containers and are equally and evenly illuminated.

Figure 2.2 *Testing for lead: (a) sampling, (b) addition of reagents, (c) formation and extraction, and (d) comparison with standards.*

The Absorption Law

The reason that the containers for samples and standards should all be of the same size is that the amount of light absorbed depends both on the length of the light path through whatever is absorbing and on the concentration. It turns out that, if the power (intensity or brightness) in a light beam is measured before and after putting an absorbing solution into it, the relationship between the two powers is:

$$P = P_0 e^{-kcb} \tag{2.1}$$

where P is the power after absorption (*i.e.* the power transmitted),

P_0 is the power before absorption (*i.e.* the incident power), e is the base of natural logarithms, k is a constant (more about this later), c is the concentration of the absorbing species in the solution, and b is the absorbing path length. All this is summarised in Figure 2.3. Equation (2.1) is one of the most powerful and widely used relationships in analytical chemistry and forms the quantitative basis of several spectroscopic techniques. The relationship is often referred to as the Beer–Lambert relationship (or law), immortalising the two scientists who first formulated it.

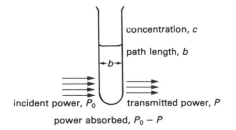

Figure 2.3 *Basic form of the Beer–Lambert law, $P = P_0 e^{-kcb}$: transmittance, $T = P/P_0$; percentage transmittance, $\%T = 100P/P_0$; absorbance, $A = \log(P_0/P) = \log(1/T) = 2 - \log(\%T)$.*

As a general rule, analytical chemists like what is measured to be directly proportional to what it is that they're trying to find, in this case concentration, c. In order to do this, equation (2.1) can be rearranged, and taking logs (to base e, ln) of both sides gives:

$$\ln(P_0/P) = kcb \qquad (2.2)$$

Converting to logs to base 10 gives:

$$\log(P_0/P) = 0.434kcb \qquad (2.3)$$

The term $\log(P_0/P)$, which is directly proportional to c, is known as the absorbance, A, of the solution. If c is expressed in mol dm^{-3}, then $0.43k$ is replaced by a new constant, ε (a Greek epsilon), so that the more usual form of equation (2.3) is:

$$A = \varepsilon cb \qquad (2.4)$$

The constant ε is known as the molar absorptivity and has the units $dm^3 mol^{-1} cm^{-1}$, as b is usually expressed in cm and A, being a logarithm, has no units (a very common error is to refer to the values as 'absorbance units'). Many spectroscopic instruments (spectrometers) are designed to measure absorbance.

What About Wavelength? Solutions of chemicals have a variety of colours because they absorb different parts of the visible spectrum (*i.e.* the wavelength range between 400 and 750 nm). Some, of course, do not absorb in the visible region and so these solutions appear colourless. A solution of the tetraoxomanganate(VII) ion (permanganate) absorbs the middle of the spectrum, allowing blue and red light to pass through, and the sensation perceived when this combination of wavelengths strikes the eye is that of the colour purple. As the eye is good at distinguishing between quite small differences in absorption wavelengths, and as the brain is quite good at remembering what colours look like, the colours of chemicals in solution provide a means of identifying them. Thus we can say that a solution possibly contains the tetraoxomanganate(VII) ion when we see the characteristic purple colour, just as our analyst friend in the pub earlier was able to distinguish beer from lager. The identification of the MnO_4^- ion becomes more positive if the colour appears as the result of some diagnostic chemistry, such as oxidation of the manganese in the solution with a strong oxidant [peroxodisulphate with silver(I) as catalyst]. Thus the absorbance of a chemical in solution is a function of wavelength. The instrument, known as a recording spectrophotometer, which measures and records this relationship is a very common piece of equipment in the analytical laboratory. The absorbance spectrum of the MnO_4^- ion, shown in Figure 2.4, is characteristic of MnO_4^-, and the identity of the species could be confirmed by comparing this spectrum with an authentic spectrum of MnO_4^-. Such a comparison of spectra is known as 'fingerprinting', for obvious reasons. The more complex a spectrum of a particular chemical is, the more use it has as a 'fingerprint'. On the other hand, however, the spectrum of a mixture of such chemicals may be difficult to interpret because of overlap of the spectra of the components.

If we want to measure how much MnO_4^- there is in the solution, it would be sensible to measure at an absorbance maximum (in this case 522 nm). As absorbance is a function of wavelength, then both ε and k are functions of wavelength. The

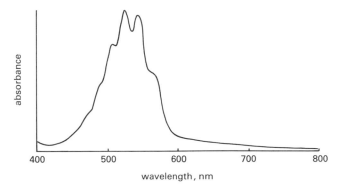

Figure 2.4 *The absorbance spectrum of MnO₄⁻ in water.*

ε value at the (or an) absorbance maximum is also a characteristic property of the chemical. If equation (2.4) is to be a valid basis of a quantitative analysis, then ε (and b) must be constant so that A is a linear function of concentration. This in turn means that the instrument must be able to make measurements over a fairly narrow range of wavelengths. So a second good reason for measuring at an absorbance maximum is that, in addition to providing the highest sensitivity, the ε values are nearly constant over the narrow range of wavelengths around the turning point. A schematic diagram of a solution spectrometer is shown in Figure 2.5. Rather than rely on values of ε and b determined by

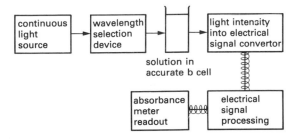

Figure 2.5 *Schematic diagram of a solution spectrophotometer. A band of wavelengths, narrow in comparison with the absorption profile of the species in solution, is transmitted by the wavelength selection device, which may be a filter but more commonly is a monochromator, containing a diffraction grating. The light intensity into electrical signal convertor is usually a photomultiplier tube (PMT).*

someone else, the normal way of performing the analysis is to prepare a number of standard solutions whose concentrations are known. The absorbances of the standards are measured and plotted against concentration to give a 'calibration curve' (of course it may actually be a straight line), and the concentrations of unknown solutions can be read from the graph (interpolation). This procedure is shown in Figure 2.6.

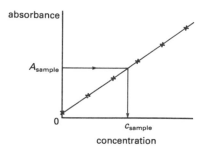

Figure 2.6   *Calibration curve for solution spectrophotometric analysis. Note that (i) the blank solution may have a finite absorbance and (ii) absorbance values range typically between 0 and 1 (100% and 10% transmittance).*

GENERAL PROCEDURE FOR INSTRUMENTAL ANALYSIS

Calibrating with standards, at the time of measuring the unknown, is a very common approach to carrying out an analysis. Although the relationship which, in theory, relates the parameter being measured to the amount of the chemical being sought may be a simple linear one, in practice interference from other components of the sample often means that instruments give different responses for different samples. In some cases the difficulties in making perfect instruments mean that there can be differences between individual instruments, and differences will arise because of different settings of the various controls. The general procedure then is to set the instrument up to give the best response for the chemical being sought, calibrate with standards, and find unknowns by interpolation. Nowadays computerised data-handling facilities mean that it is not necessary actually to draw the graph shown in Figure 2.6. The computer will fit an equation to the points and calculate unknown concentrations.

WHY DO CHEMICALS ABSORB RADIATION?

Theories and Models

There is no way of providing an absolutely true answer to the above question, which involves the way radiation interacts with matter, because we only have theories about what radiation and matter are. However, physicists and chemists have amassed an enormous amount of experimental evidence in support of our present theories. The experimental evidence about radiation is very interesting. In some experiments radiation behaves as though it consists of an oscillating electric field with an oscillating magnetic field at right angles to it, and in some experiments radiation behaves as though it consists of discrete packets of energy (each packet being known as a photon). The experimental evidence concerning atoms supports the theory that each consists of a small, positively charged nucleus (in which most of the atomic mass is concentrated) surrounded by negatively charged electrons. The electrons move in regions of a particular shape and energy depending on the number of electrons in the atom. These regions are known as orbitals, and their shape is related to the probability of finding an electron in the particular region. Each orbital in each atom has a definite energy associated with it. In addition to moving relative to the nucleus the electrons have another sort of motion called spin.

Theories are closely related to models. The model may be a mathematical one in that the observed behaviour may be 'explained' by an equation (such as the Schrödinger equation describing the spatial distribution of the probability of finding electrons) or it may be a physical one (such as thinking of an atom as being like a small 'solar system' with small spherical electrons orbiting a central nucleus). Models, especially physical ones, are very useful in allowing us to think about and visualise processes on the atomic and molecular scale. But it should not be forgotten that they are only models and are limited by their use of concepts and terms taken from our everyday life, where objects have sharp, well defined boundaries and move in ways we can readily predict from our previous experience. At the atomic and molecular level things are much less clear-cut and quantities like energy are no longer continuous functions but are quantised, *i.e.* only have certain fixed values. Thinking about the behaviour of atoms and molecules usually involves the selection of the most appropriate model for the situation being considered.

Absorption of Radiation

The energy, E, of a light photon is proportional to the frequency of the associated wave motion, ν, the constant of proportionality being Planck's constant, h (6.626×10^{-34} J s). The frequency is inversely proportional to the wavelength, λ, the constant being the speed of light, c (2.998×10^8 m s^{-1}). Thus:

$$E = h\nu = hc/\lambda \qquad (2.5)$$

If a chemical species possesses vacant orbitals which can accept electrons from lower-energy orbitals, it may be possible to induce an electronic transition by irradiating the chemical with radiation of the appropriate wavelength – appropriate, that is, to the energy difference between the two orbitals. The photon model for light and an energy level diagram for electrons based on the orbital model provide a suitable basis for an overall model of the situation. The light beam consists of a stream of photons, each of energy $E = hc/\lambda$, some of which collide with electrons and transfer all their energy to them. If this energy is exactly the value of the energy difference between the occupied orbital and the vacant higher-energy orbital, the electron may be transferred to the higher-energy orbital. Some of the photons do not collide with electrons, but obviously the greater the number of appropriate electrons in the photon light path the higher the probability of absorption occurring. Thus the reason for absorbance increasing with path length and concentration can readily be appreciated. A detailed mathematical consideration of the statistics of collisions produces equation (2.1). The model used here, summarised in Figure 2.7, is not adequate to explain all absorption phenomena, as some electronic transitions which might be expected to occur on the basis of occupancy and energy considerations are not observed. To 'explain' these observations, the way the orbitals overlap on a quantum mechanical basis has to be considered.

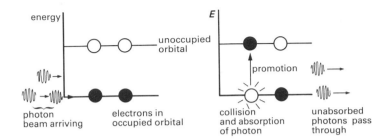

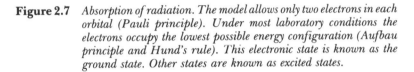

Figure 2.7 *Absorption of radiation. The model allows only two electrons in each orbital (Pauli principle). Under most laboratory conditions the electrons occupy the lowest possible energy configuration (Aufbau principle and Hund's rule). This electronic state is known as the ground state. Other states are known as excited states.*

Absorption and Structure

The energy differences between the outer orbital and the various unoccupied orbitals of many atoms, ions, and molecules (both 'free' in the gas phase or in solution) correspond to wavelengths in the near ultraviolet (UV) and visible parts of the spectrum. Absorbance measurements over the range 200—800 nm are very commonly made with the same instrument, known as a UV–visible spectrometer, which usually contains two light sources to produce radiation over this range of wavelengths. The energies of the various orbitals in atoms, molecules, and ions are a unique property of the particular chemical species, as they depend on the number of electrons, nuclear charge, and relative location of nuclear charges (in the case of polyatomic species), so the absorption spectrum is a unique property of that chemical species. This was discussed earlier in relation to the identification of the MnO_4^- ion in solution. In principle, UV and visible absorption spectrometry should tell us something about the electronic structure of atoms, molecules, and ions. In fact the technique is most powerful for monoatomic species in the gas phase. Vibrational effects (as shall be discussed shortly) and interactions with solvent molecules make this type of spectroscopy less useful for finding out the electronic structure of polyatomic species in solution. However, changes in the absorption spectra of polyatomic species due to changes in molecular structure are very widely applied in chemical analysis.

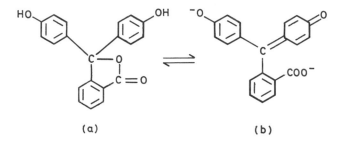

Figure 2.8 *Acid form* (a) *and basic form* (b) *of a potential acid–base indicator molecule.*

As a first example consider the molecule shown in Figure 2.8(a). In acid solution this molecule is colourless (it absorbs only in the UV because of the phenyl rings); however, if the solution becomes alkaline, the structure changes to that shown in Figure 2.8(b). Now there is a molecular orbital spread over the whole molecule and the energy spacing between the ground state and the first excited state has decreased so that it corresponds to the blue end of the visible spectrum. The resulting effect is that the solution appears red. This behaviour might provide the basis for the use of this molecule as an acid–base indicator. The spectroscopic behaviour is not the only criterion that has to be satisfied: the equilibrium constant for the acid–base indicator reaction must have an appropriate relation to the main analytical acid–base reaction. This aspect will be discussed later in Chapter 4. The design of indicators for other types of titrations (redox, complexometric) must be based on similar considerations as far as the colour changes are concerned.

As a second example consider the general approach to the spectrophotometric detection of metals. Solutions of most metals are either colourless or faintly coloured, particularly at the $mg\ dm^{-3}$ level, where it is very important to be able to make reliable concentration measurements. The general approach is to react the metal, M, with a reagent, L (normally a complexing ligand), to form a coloured complex ML_n. If the product is not separated, its molar absorptivity at the wavelength chosen for measurement must be greater than the molar absorptivity of the reagent L; otherwise in the presence of the metal no increase in absorbance will be observed. The desired effect is illustrated in Figure 2.9 for a method of determining tin. The molar absorptiv-

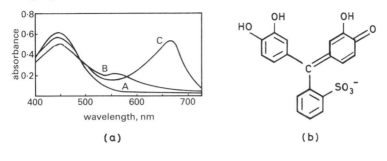

(a)

(b)

$$C_{10}H_{33}(CH_3)_3\overset{+}{N}Br^-$$

(c)

Figure 2.9 (a) *The absorbance spectra of* A, *the reagent catechol violet,* B, *the complex between catechol violet and tin(IV), and* C, *the complex formed on addition of cetyltrimethylammonium bromide (CTAB). Such a shift in absorbance to longer wavelength is known as a bathochromic shift.* (b) *Catechol violet.* (c) *CTAB.*

(Reproduced with permission from R.M. Dagnall, T.S. West, and P. Young, *Analyst*, 1967, **92**, 27.)

ity of the tin–catechol violet complex at 555 nm is 65 000 dm³ mol⁻¹ cm⁻¹, but on addition of cetyltrimethylammonium bromide a further shift in absorbance maximum occurs to 620 nm and the molar absorptivity increases to 95 600 dm³ mol⁻¹ cm⁻¹. As an absorbance value of 0.01 can easily be measured (and seen, if the wavelength is in the middle of the visible range, where the eye is most sensitive), equation (2.4) predicts that in a 1 cm cell about 0.01 mg dm⁻³ should be detectable. However, the discoverers of this reaction reported that the lower limit of the validity of equation (2.4) was 0.2 mg dm⁻³. There are a number of possible reasons for this. In order to drive the ML_n formation reaction as far as possible to completion, an excess of L would be used, producing a finite absorbance at 620 nm. The reagent (and other solutions) may be difficult to purify and may contain sufficient metallic impurities to give an absorbance much greater than 0.01. The reaction conditions are adjusted to pH 2.2 from a strongly acid solution by the addition of dilute ammonia solution. The original acid and the ammonia solutions could both contain trace metals.

If such a large shift in wavelength cannot be achieved and the excess ligand still has a considerable absorbance at the wavelength to be used, it may be necessary to perform some chemistry on the system to remove the excess reagent. This is done when dithizone is used as the spectrophotometric reagent. After formation of the metal dithizonate in tetrachloromethane (see page 20) the excess dithizone reagent is removed from the organic layer by shaking with an aqueous solution of about pH 10, metal dithizonates being stable at this pH.

The use of solution spectrophotometric methods for determining trace metals has declined somewhat over the past 10—20 years or so (as far as laboratory-based determinations are concerned) because of the widespread use of another technique known as atomic absorption spectrometry, which is described in the next section. However, there are many spectrophotometric methods in use for trace anionic species.

UV (and occasionally visible) absorption methods are widely used for the quantitative determination of organic compounds, particularly in conjunction with a separation technique such as chromatography (see Chapter 6). The techniques can also provide some information about the structure of unknown organic compounds.

ATOMIC ABSORPTION SPECTROMETRY (AAS)

Analytical chemists make extensive use of the absorption of radiation by free atoms as the basis of an analytical technique for measuring $mg\ dm^{-3}$ concentrations of metal ions. The absorbance values measured for a given sample are much more dependent on the individual spectrometer used than is the case for molecular absorption spectrometry, and so analytical chemists need to understand how atomic absorption spectrometers work.

There are two practical problems to be overcome in applying atomic absorption as a laboratory analytical technique. The first is that the sample will almost certainly not consist of free atoms. The energy required to produce atoms is usually provided by chemical reaction, namely the oxidation of acetylene. The reactants are air and acetylene, producing a lower-temperature and slower-burning flame than the rather unmanageable oxyacetylene flame (used for welding, *etc.*). The gases are premixed and then fed to a burner designed to produce a long (10 cm typically), thin (2—3 mm) flame. The sample is dissolved and introduced into the gas mixing chamber as a fine spray. What happens then is summarised in Figure 2.10. A small proportion of the original material

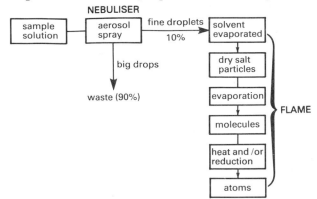

Figure 2.10 *Summary of physical and chemical processes leading to atom formation in a combustion flame. The production of the aerosol is called nebulisation.*

finally ends up as atoms in the flame. Provided there is a steady uptake of solution by the nebuliser, the rate of atom production becomes constant and a steady absorbance value may be measured. Some elements form very stable molecular species in the flame (such as oxides or mixed oxides with elements such as aluminium, silicon, or phosphorus), and a higher-temperature flame consisting of dinitrogen oxide (N_2O) and acetylene is used This produces temperatures of about 3000 °C whereas the air–acetylene flame temperature is about 2000 °C.

The second problem is that atoms only absorb over a very narrow range of wavelengths – maybe only 0.01 nm – unlike the many tens of nm over which molecular species in solution absorb (see for example Figures 2.3 and 2.9). It is not possible to use the type of spectrometer shown in Figure 2.5 with the solution replaced by a flame to detect atomic absorption. The reason for this is that the range of wavelengths that can be selected from such a 'continuum' source (either an electrical discharge through deuterium for the UV or an incandescent tungsten filament for the visible) and still give enough intensity to be measured is very much larger than the range over which the atoms are absorbing. The small amount of radiation absorbed by the atoms is just not detected by the device which converts light intensity into an electrical signal.

What is needed, just as for the molecular case (see Figure 2.5), is the light source's emission wavelength profile to be much narrower than the atom's absorption wavelength profile. This is achieved by using a lamp specially designed to produce not only such a narrower emission profile but also the profile centred at exactly the right wavelength for the appropriate atoms to absorb. The special light source is called a hollow cathode lamp, because one of the electrodes is made in the shape of a hollow cylinder and lined with the metal to be measured.

When a high voltage is applied between the cathode and a metal rod anode, both sealed into a glass cylinder containing low-pressure neon, the neon ionises and the impact of positively charged neon ions on the inside of the cathode is sufficient to vaporise some of the cathode lining as free atoms. Further collisions are sufficiently energetic to excite electrons in some of these atoms to higher orbitals, and when these electrons return to the original lower-energy orbitals light is given out. This atomic emission is at exactly the right wavelength to be absorbed by the identical atoms in the flame. The whole process is summarised in Figure 2.11. The width of an emission or absorption wavelength profile increases with temperature and pressure. In addition, the width of an emission profile depends on its temperature profile and physical size (the length and temperature gradient in the

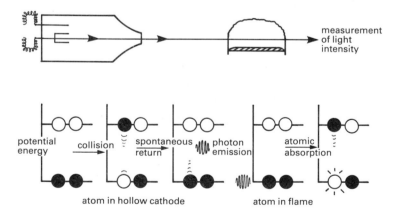

Figure 2.11 *The hollow cathode light source for atomic absorption. As the element emitting in the lamp is the same as that being measured in the flame, the electron configurations are identical and the energy level diagrams are identical; hence the wavelength emitted from the lamp will be absorbed in the flame.*

direction of observation being most important). The hollow cathode lamp is deliberately designed to reduce the emission profile width compared with the absorption profile width by being a low-temperature (a few hundred °C), low-pressure device. The source is compact and at uniform temperature. Thus the necessary condition for making absorbance measurements is achieved.

A different lamp is required for each element, but of course each lamp is specific for each element since no other atoms, produced from other components of the sample, will absorb the radiation. For example, it is possible to determine nickel in steel simply by dissolving the samples and spraying the solution into the flame. The presence of other elements such as cobalt, chromium, manganese, and iron (large excess, of course) does not cause any interference, as atoms of these elements will not absorb the light from a nickel hollow cathode lamp. Interferences do arise, however, from anything that affects the sequence of atom-forming steps shown in Figure 2.10 or the intensity of the light measured as shown in Figure 2.12. Such interferences are very common, and the analytical chemist using or devising an atomic absorption method of analysis must understand how these arise and how they may be dealt with.

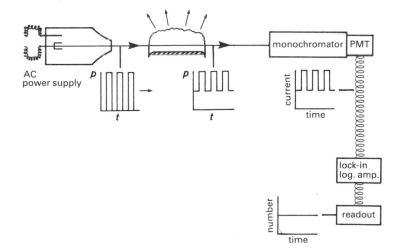

Figure 2.12 *Atomic absorption spectrophotometer with modulated source and lock-in amplification.*

As with other spectroscopic methods, some interference effects can be overcome by the use of chemistry and some by the appropriate design features of the instrument (see Box 1). Atomic absorption spectrometers must be calibrated at the time of making the analysis, and the analytical chemist must think carefully about the calibration procedure to be adopted (see Chapter 5).

Applications

Despite the various interference effects, atomic absorption spectrometry is a very widely used technique for trace metals (mg dm^{-3} level in the nebulised solution) in a whole range of sample types, including food, drink, blood, wine, waters (tap, river, surface run-off, sea, *etc.*), alloys, drugs, paint, plastics, rocks, oils, plants, soils, and anything else you can think of.

The scope of the technique can be extended to even lower concentrations by using a means of producing a more concentrated atomic vapour. One way of doing this is to replace the flame with a small carbon tube furnace which may be heated up to 3000 °C by electrical resistance heating. The light from the hollow cathode lamp passes down the middle of the tube furnace.

OTHER TYPES OF OUTER-ELECTRONIC SPECTROMETRY

So far we have looked at the absorption of radiation by molecules in solution and atoms in flames (or furnaces) as analytical methods based on the transition of an outer electron from a filled or partially filled orbital to an unoccupied higher orbital. However, as we've already seen for the atoms in the hollow cathode lamp and the combustion flame, it is possible for the reverse process to occur by emission of a photon. There are a whole range of analytical spectrometries based on this process for both atoms and molecules. Electronic emission methods are known in general as luminescence methods (although the term is more often used for molecules than atoms) and are classified by the way in which the excited state is produced. The particular method which uses light to produce the excited state, by absorption of photons, is known as fluorescence (usually short lived) or phosphorescence (usually long lived). The continuum emission obtained from hot

BOX 1 Interference Effects in Atomic Absorption Spectrometry

A commonly occurring interference is stable-compound formation in the flame. If a calcium solution containing aluminium is nebulised, then the number of atoms formed is less than if the same concentration of calcium alone were nebulised. This is due to the formation of a thermally stable calcium aluminate in the flame (often such compounds have no definite composition and are written as 'Ca–O–Al'). This interference may be overcome by adding to the solution an element which forms a more stable aluminate than does calcium. Such an element is lanthanum, which, added in excess to help the equilibrium Ca–O–Al + La $\rightleftharpoons$ Ca + La–O–Al over to the right, produces calcium as free atoms. The lanthanum (added as the chloride) is known as a releasing agent.

Another type of interference is something that may have struck you already: if collisions with atoms in the hollow cathode lamp can produce excitation and atomic emission, then the same process should occur in the flame. The light intensity measured will then be too high because it will be that due to the hollow cathode lamp emission minus that absorbed in flame plus that emitted by the flame. In fact the flame emits at several wavelengths owing to the chemiluminescent flame gas reactions (the characteristic blue of the two combustion zones of a hydrocarbon–air flame) and the hollow cathode emits wavelengths corresponding to atomic emission from the fill gas (the red neon emission). The first stage of the instrument design is to put a monochromator between the flame and the photomultiplier tube (contrast this with the arrangement shown in Figure 2.5) so that only light in the wavelength range in the immediate vicinity of the desired wavelength reaches the PMT. The second stage in the design is to discriminate against the steady atomic emission from the flame by using an alternating (often referred to as a.c.) power supply for the hollow cathode lamp. This produces a discontinuous emission that is rapidly switched on and off (you don't see this with the naked eye unless the frequency of switching is quite low: light bulbs run from the 50 Hz mains supply appear to emit continuously). This process is known as modulation, and the signal-processing circuitry is designed so that it only passes a.c. signals of the same modulation frequency to the readout device.

The light intensity falling on the PMT consists of two types, a steady or d.c. emission from the flame and an a.c. component from the hollow cathode lamp, decreased from its original intensity by the absorption in the flame. The resulting electrical signal has the same characteristic, but the d.c. component is removed by the 'lock-in' nature of the amplifier. The amplitude of the a.c. signal corresponds to P in equation (2.3), P_0 being measured when distilled water is nebulised. The electrical circuitry will be designed to calculate $\log(P_0/P)$ and to display this number on the readout device. The whole process is shown in Figure 2.12.

bodies is known as 'incandescence'. It does not provide any useful analytical information as the characteristics of the light emitted are not related closely to chemical composition.

Atomic Emission Methods

The analytical basis for all of these spectroscopic techniques is that the intensity of emitted radiation is directly proportional to the number of emitting atoms, which in turn is directly proportional to the total number of atoms, in turn proportional to the concentration of the element in the sample. The intensity of emitted radiation increases very rapidly with increasing temperature, so in general the hotter the atom source the brighter the emission. The most common of these methods is flame emission spectrometry, particularly for elements whose emission is fairly easy to excite, such as lithium, sodium, potassium, rubidium, and caesium. Sodium and potassium are commonly measured, for example, in urine and blood plasma in clinical laboratories by simple instruments called flame photometers.

A larger amount of energy can be transferred to a solid sample by striking an electric arc or spark between the sample and a carbon electrode. To produce such a discharge at atmospheric pressure may require many thousands of volts (as opposed to the few hundred needed for the hollow cathode lamp) and the temperature in the arc or spark gap is very high (anything between 4000 and 8000 °C). As a result many of the higher electronic energy levels can be reached and some elements (particularly transition elements) can emit at a large number of wavelengths. A very good-quality wavelength separation device is required, as all the elements in the sample will be atomised and excited to some extent. Thus it is possible to perform simultaneous multi-element analysis if the instrument uses a photographic detection method or incorporates a number of detectors (photomultiplier tubes). This latter technique, known as 'direct-reading emission spectrography', is often used in the metallurgical industries where rapid analysis may be very important. To produce steel of the required specification, the concentrations of a number of alloying elements must be within the appropriate concentration ranges; it would prove very expensive to cool the steel and then find the composition to be out of specification. So, although it could cost thousands of pounds to keep the steel in a converter molten while a sample is taken away for analysis, a rapid analysis at this stage is the most cost-effective option.

One of the most recent developments in atomic emission spectrometry uses an inductively coupled plasma to atomise and excite the sample. With this device temperatures of up to 10 000 °C are possible, by using an oscillating radiofrequency electric field to induce rapid, oscillating movements in electrons and ions. The plasma is first formed in ionised argon and the sample solution introduced as an aerosol. There is so much energy that the sample material is completely atomised and there are no interference effects due to chemical reactions.

Another recent development in analytical atomic emission spectrometry is to measure atomic fluorescence, the light source being a hollow cathode lamp and the atoms being produced in an argon plasma. Because of the selective excitation mechanism, the detector can be 'locked in' to the modulation frequency of the lamp, and thus the unwanted atomic emission from the plasma may be ignored, just as is done in atomic absorption spectrometry to discriminate against light emitted from the flame (see Box 1).

Molecular Emission Methods

The most commonly used method in this category is fluorescence in solution. The analytical basis of this is given in Box 2.

Fluorescence methods are more sensitive than absorption methods as the instrument measures emitted radiation against a fairly dark background. Normally fluorescence is measured at right angles to the incident light, to minimise the amount of scattered light reaching the detector. There will also be a wavelength selection device (monochromator) on the emission side as well as on the excitation side of the sample. Applications of fluorescence spectrometry are not as widespread as those of molecular absorption spectrometry, but the use of the technique is increasing. Although many different molecules absorb radiation, very few emit fluorescent (or phosphorescent) radiation, and it is often necessary to perform some chemistry on the sample in order to form a fluorescent derivative of the particular species being sought. Compounds containing NH_2 groups, such as amino acids, can be converted to fluorescent molecules by reaction with a molecule called fluorescamine. The excess reagent is hydrolysed to a non-fluorescent product. When a small

BOX 2   The Analytical Basis of Fluorescence Spectrometry

The brightness of measured fluorescent radiation, P_f, emitted after absorption of light energy will depend first of all on the amount of light absorbed, *i.e.* in the notation used for equation (2.1) (page 22) P_f will be proportional to $P_0 - P$ (see Figure 2.3). However, not every photon absorbed will give rise to an emitted photon as some of the excited-state molecules lose energy by collision (most likely with solvent molecules as these are present in the highest concentration). This is accounted for by including a factor, ϕ, known as the quantum efficiency in the final equation. Finally, although the fluorescent radiation is emitted in all directions, the fluorimeter (as a fluorescence spectrometer is known) will only measure the fluorescence in a restricted range of directions. A further factor, k', the geometric factor, is included in the final equation, which can now be written as:

$$P_f = \phi k'(P_0 - P) \qquad (2.6)$$

We now want to find how P_f is related to concentration, c, so we substitute for $P_0 - P$ the expression obtained by subtracting both sides of equation (2.1) from P_0, namely:

$$P_0 - P = P_0 - P_0 e^{-kcb} \qquad (2.7)$$

$$\therefore P_f = \phi k' P_0 (1 - e^{-kcb}) \qquad (2.8)$$

Analytical chemists want to use techniques that have a linear relationship between the measured quantity and concentration, but this is not given by equation (2.8). To get such a relationship from equation (2.8) an approximation has to be made. This is based on the expansion:

$$e^{-kcb} = 1 - kcb + [(kcb)^2/2!] - [(kcb)^3/3!]\ldots \qquad (2.9)$$

and on neglecting squared and higher-order terms on the basis that the value of kcb will be small. Thus:

$$P_f = \phi k' P_0 kcb \qquad (2.10)$$

Substituting in equation (2.10) for k in terms of ε (see page 23) produces the usual form of the equation:

$$P_f = \phi k' \times 2.303 \varepsilon cb \qquad (2.11)$$

The usual restriction on the validity of equation (2.11) is taken to be $\varepsilon cb < 0.05$ [εcb is the absorbance of the solution – see equation (2.4)]. When εcb equals 0.05, equation (2.11) is in error by about 5%.

Fluorimetry is only one case of a phenomenon exploited for analytical purposes for which a suitable basic equation is only obtained by introducing approximations. For example, the equation for atomic absorption spectrometry which linearly relates absorbance to concentration depends on several approximations in its derivation. Analytical chemists need to be aware of these approximations and of the limitations they impose on the use of the techniques in question.

fluorescent molecule is tacked on to a much bigger molecule, this is referred to as fluorescence 'labelling'. Fluorescence methods are also being used in the identification of the source of illegal dumping of crude oil into the sea. Such crudes and the weathered oil washed up on beaches can be matched because the complex mixture of hydrocarbons often gives rise to a characteristic set of fluorescence spectra.

It is also possible, for a small range of molecules, to produce light emission from an excited state produced by a chemical reaction. This process is known as chemiluminescence and is capable of detecting very small concentrations indeed. For example, as little as 10^{-3} ng of cobalt can be measured by its catalytic effect on the chemiluminescent oxidation of N,N-3-aminophthaloylhydrazine (luminol) by hydrogen peroxide (see Figure 2.13).

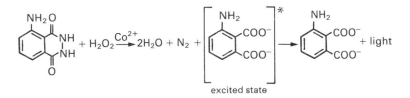

Figure 2.13 *Cobalt-catalysed oxidation of luminol by hydrogen peroxide.*

As with fluorescence, there is a growing interest in finding new reactions with the appropriate chemiluminescent properties to form the bases of new analytical methods. Most of this interest centres around chemical reactions in solution, but it is possible to produce chemiluminescence from reactions in a flame. The hydrogen flame appears to be particularly suitable, and a range of non-metallic anionic species can be measured. For example, sulphur-containing anions give rise to blue emission from S_2 and phosphorus anions produce a green emission from HPO. The emission intensity can be considerably increased if the sample is introduced into the flame in a small cavity. This technique is known as MECA, molecular emission cavity analysis.

OTHER ABSORPTION SPECTROMETRIES

So far in this chapter spectroscopic methods based on the interaction of light with the outer electrons in atoms and molecules have been described. This is by no means the total extent of analytical spectroscopy, and analytical chemists (and others) make extensive use of some other types of absorption spectrometry.

Infrared Absorption Spectrometry

The region of the electromagnetic spectrum on the low-energy side of the visible is known as the infrared (IR) region. It extends from about 800 nm to 1 000 000 nm, and, although analytical information can be obtained from absorption of radiation throughout this wavelength range, the most useful part is a middle portion extending from about 2500 to 16 000 nm or 2.5 to 16 μm (micrometres). Most laboratory instruments are constructed to cover this wavelength range.

The model that is used to explain the absorption of IR radiation is different from the photon collision model used to explain absorption of UV and visible radiation. In the case of IR radiation it is the quantised energy levels associated with vibrations between atoms within the molecule that are involved. How then can a vibrating molecule absorb energy from a light beam? Well, in general, when atoms vibrate in a molecule, the 'centre of gravity' of the positive charge (on the nuclei) does not coincide with the 'centre of gravity' of the negative charge (the electrons) and so the molecule contains an oscillating electric dipole. The way in which this dipole oscillates will be quite complicated, but it can be resolved into a number of simpler motions due to the individual vibrations of small groups of atoms. The simplest of these is the stretching and contraction of a bond between two atoms, but groups of atoms will be rocking, wagging, scissoring, and twisting, and with two atoms linked to a central atom the stretching may be symmetric or asymmetric. If the frequency of an individual vibration's oscillating dipole is the same as that of the oscillating electric dipole of the light beam irradiating the sample, then energy may be transferred from the light beam to the vibrating atoms. This decreases the amplitude of the beam (hence its intensity, which is proportional to the amplitude squared) and increases the amplitude of the molecular vibration. The process is called resonance. It is the spectroscopic analogue of the process by which a guitar string may be made to vibrate when a neighbouring string, fretted to give the same note, is plucked. We can also use our everyday ideas about the relative frequencies of vibrations to predict that the stronger the bond and the lighter the atoms the faster the vibration will be, whereas the weaker the bond and the heavier the atoms the slower the vibration.

So the photon collisional model that was used for the absorption of UV and visible radiation in which light intensity was considered to be the rate of arrival of photons cannot readily be applied to the phenomenon of absorption of IR radiation. Here the wave nature model of radiation is used and light intensity is considered to be proportional to the square of the amplitude of the oscillation of the electric field. If a vibrating electric dipole can absorb energy by resonance, then the intensity of the light beam will be reduced.

The absorption law [equation (2.4)] still holds, and absorbance is directly proportional to concentration. However, the molar absorptivity values are much lower than for UV or visible absorption, and IR absorption cannot be used for trace analysis except for a few strong absorbers which may be measured with specially designed spectrometers with very long path lengths. The main use of IR spectrometry is for identifying molecules and deciphering their structure (see Box 3).

BOX 3 Structure Analysis by Infrared Spectrometry

A non-linear molecule containing n atoms has $3n - 6$ ways of vibrating (called normal modes of vibration). Some of the vibrations may combine to give new vibrations which can also absorb, some vibrations will be at overtone frequencies (simple integral multiples of the lowest or fundamental frequency), and the overtone vibrations may combine as well. The net result is that the IR absorption spectrum can be quite complicated with many overlapping bands but is thus characteristic of a particular molecule. The IR spectrum is a molecular 'fingerprint' and is used by analytical chemists for qualitative analysis.

In addition, it turns out that certain groups of atoms in a molecule always vibrate at about the same frequency but that the exact frequency depends on the other components of the molecule. For example, the stretching frequency of the carbonyl group, $C=O$, occurs at around 5.1×10^{13} Hz, *i.e.* just under 6000 nm. This can vary from 5.14×10^{13} Hz for an aromatic ester to 5.31×10^{13} Hz for a monocarboxylic acid. The same sort of effect is observed with other groups such as CH, OH, $C=C$, and $C\equiv N$, so it is possible to correlate absorption bands in the spectrum with particular functional groups in the molecule and thus to be able to identify parts of the structure of the molecule.

The infrared spectrum of n-propyl ethanoate is shown in Figure 2.14

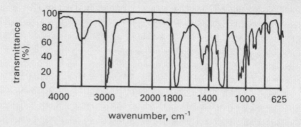

Figure 2.14 *Infrared absorption spectrum of n-propyl ethanoate.*

in the usual way a laboratory spectrometer displays a spectrum. This should be contrasted with the way a UV or visible absorption spectrum is normally displayed [see Figure 2.9(a)]. Note that the vertical scale is the other way up, with zero absorption at the top, and it is calibrated in terms of percentage transmission ($100 \times P/P_0$). Also note that the horizontal scale is calibrated neither in frequency nor in wavelength units but in reciprocal length. As can be seen from equation (2.5), reciprocal length is directly proportional to energy, the constant of proportionality being hc, and this may be one reason why this method of designating the energy of this part of the spectrum is used. If reciprocal centimetres (cm^{-1}) are used, the numbers produced are not too cumbersome to handle, as the spectral range of 2500 to 16 000 nm corresponds to 4000 to 625 cm^{-1}. This designation of reciprocal length is called 'wavenumber'. The corresponding frequencies are 1.2×10^{14} to 1.875×10^{13} Hz. Thus wavenumbers increase with increasing frequency, so that the higher the wavenumber the stronger the bond or the lighter the atoms.

As with atomic absorption spectrometry, there are some practical problems that the instrument is designed to surmount before a useful measurement may be made. Analytical chemists must understand what these are and how the instrument works if the best use is to be made of the technique. The major problem is that the air inside the instrument contains two components which absorb IR radiation quite strongly, namely carbon dioxide and water vapour. In order that the absorption of the CO_2 and H_2O is not measured by the instrument, the 'double-beam' design is used; this is explained in Figure 2.15. In addition, the double-beam arrangement will compensate for any variation in light source emission intensity and any variation in detector response, both of which are functions of wavelength and thus would distort the spectrum (scan of percentage transmission versus wavenumber) produced. These latter problems are not confined to the IR region but also have to be overcome in the UV and visible regions. Figures 2.14 and 2.9(a) have been obtained with double-beam instruments.

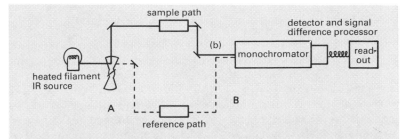

Figure 2.15 *The double-beam principle. Light from the source is alternately directed around two identical optical paths by the rotating mirror sector A; the paths are recombined by a second rotating sector at B. The signal-processing electronics are designed to measure the difference in intensity between the two beams. As CO_2 and H_2O will absorb equal amounts from each beam, the net result will be zero absorption at the readout. Only when something in the sample beam absorbs differently from something in the reference beam will a signal appear at the readout.*

Nuclear Magnetic Resonance Spectrometry

Although not very widely used for quantitative analysis, nuclear magnetic resonance (NMR) spectrometry is a very powerful method for sorting out the molecular structure of an organic compound. The basis for the technique is a little bit more complicated than for the other kinds of spectrometry we have looked at so far, and we have to use a model that is a bit further removed from everyday situations than the ones we used for UV, visible, or IR absorption. Firstly, the effect is only observable in very strong magnetic fields, so the instrument has to supply this around the sample; secondly, it is the magnetic properties of the nucleus that give rise to the effect. The basic principles of the technique are explained in Box 4. The magnetic field strengths used are such that absorptions in the radiowave region of the spectrum are obtained.

An example of a proton (^{1}H) NMR spectrum is given in Figure 2.16 and a ^{13}C spectrum in Figure 2.17. The area under the peaks is directly proportional to the number of nuclei in that particular environment. Because the technique is so useful, a very large amount of effort has been devoted to developing improved spectrometers, with the result that the best-quality results are only

BOX 4   Basic Principles of Nuclear Magnetic Resonance

Some nuclei (depending on the number of protons and neutrons) are considered to be spinning. Thus they can be thought of as moving electric charges and will therefore have an associated magnetic moment. In the presence of an external magnetic field the magnetic moments will precess (like a gyroscope in the earth's gravitational field) at a fixed frequency about an axis which lines up with the external magnetic field direction. However, the orientation of the precession can only be in certain directions relative to the field direction. In the simplest case there are two possibilities: the nuclear 'magnet' lines up either with the field or against the field. These two configurations differ in energy, and, by applying energy from a beam of electromagnetic radiation of the appropriate frequency, the orientation of the precession can be changed from the lower-energy state (with the field) to the higher-energy state (against the field). In this case the precessing nuclear magnet has absorbed energy from the oscillating magnetic field of the radiation, decreasing its intensity, by a process of resonance similar to that described for the absorption of IR radiation by an oscillating electric dipole.

Not all nuclei possess a magnetic moment, but among those that do ^{1}H and ^{13}C are the most important. Both of these nuclei have the simplest type of magnetic moment which gives rise to only two orientations in a magnetic field. The energy difference between the two states depends on the strength of the external magnetic field, and for strengths of 14.09 kGauss (this is about 80 000 times stronger than the Earth's magnetic field and poses quite a few problems for the instrument manufacturers) the resonant frequency is 60 MHz, *i.e.* radiofrequency electromagnetic radiation is required.

From the explanation so far it appears that all protons (for example) in a molecule will resonate at the same frequency. However, the beauty of the technique is that the external magnetic field induces circular motion of the electrons in the molecule, which in turn produces a localised magnetic field opposing the applied field. The extent of this shielding (as it is called) depends on the electron density around the proton, which in turn depends on the bonding. A different chemical environment means a different local magnetic field strength and thus a different resonant frequency. Although the effect is very small, this chemical shift is the basis of the use of NMR for structural analysis. It is normally expressed as the shift in frequency from the resonant frequency of a universally agreed reference compound, tetramethylsilane, $(CH_3)_4Si$, divided by the instrument operating frequency, and multiplied by 10^6. The dimensionless number produced (given the symbol δ) is referred to as 'p.p.m.' (parts per million) and has values of between 0 and 15 for protons and 0 and 200 for ^{13}C nuclei.

Interpreting an NMR spectrum requires a bit of practice, because the spectrum is complicated by the fact that the nuclear spins interact with each other, so the peak corresponding to the resonant condition for a particular chemical environment is split into a number of components. This splitting depends on how many other spinning nuclei are nearby.

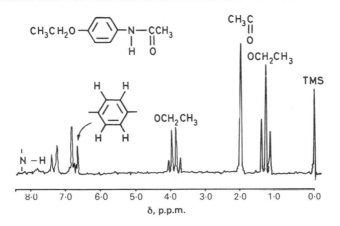

Figure 2.16 *Proton NMR spectrum of p-ethoxyacetanilide. The vertical scale is absorption.*

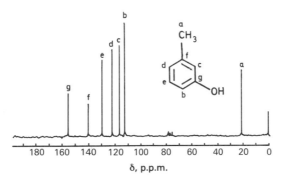

Figure 2.17 *^{13}C NMR spectrum of m-hydroxytoluene. This was recorded using an instrument which decoupled the interaction between the carbon and hydrogen nuclei. Interaction between the ^{13}C nuclei is not observed because of the low abundance of this isotope. The spectra are obtained by averaging repetitive scans.*

obtained from very expensive pieces of equipment of such operational complexity that a specially trained operator is employed to keep the instrument in good working order and to run the spectra of the compounds being investigated.

OTHER TYPES OF SPECTROSCOPY

You will recall that in this discussion of how the analytical chemist makes light work we started in the visible region of the electromagnetic spectrum and, after considering the UV region as well, moved into the IR and then the radiowave regions. This is by no means the whole story. In the microwave region, for example, electron spin resonance (very similar in concept to NMR) provides information about unpaired electron spins in a compound, and information about structure can be obtained from the absorption due to molecular rotation, which is a quantised phenomenon just as vibrations are.

At the other end of the energy scale, there are a number of spectrometries based on the use of X-rays. Despite the possible hazards of such ionising radiation, X-ray spectrometries are quite widely used. The diffraction of X-rays is used to provide information about the structure of crystals, whereas X-ray emission, usually as the final step in the fluorescence effect, is used for both qualitative and quantitative analysis. X-Ray fluorescence is particularly useful for solid samples, as X-rays will penetrate quite deeply into the solid, removing core electrons as they are absorbed. The resulting vacancies in the inner electronic orbitals are filled by electrons from less tightly bound orbitals. The resulting excess energy possessed by the ion may be liberated as an X-ray photon. This can escape from the sample and be measured. A large number of elements can be detected and measured since the core binding energies for electrons in the same orbitals vary considerably from one atom to another, although the technique is not as good for light elements as for heavy ones. Under favourable conditions, as little as 10 mg kg^{-3} of an element may be detected. The method is rapid and non-destructive.

At even higher energies, gamma radiation (another potentially harmful ionising radiation) is used as the basis of certain analytical methods. Gamma rays are produced as the result of certain types of reactions within the nucleus. Usually the sample is made radioactive artificially by irradiating it with neutrons, then the γ-ray spectrum is recorded and analysed. The energy and intensity are related, respectively, to the particular element (qualitative analysis) and to how much of it is present (quantitative analysis). Obviously a specialised laboratory is required for such neutron activation analysis, as the technique is known, but several such laboratories exist that provide an analytical service to anyone who is prepared to pay the going rate for the job. These methods are very accurate indeed as, unlike a lot of other analytical methods, the results do not depend in any way on the chemical environment of the element. On the other hand, the method is totally inappropriate for an analytical situation in which the particular chemical form of the element is the important factor.

THE PARTICLE SPECTROMETRIES

All the methods that have been discussed so far have been based on a study of the interaction of electromagnetic radiation with atoms, molecules, *etc*. There are a number of techniques which provide information from the analogous study of the particles that the sample emits under certain conditions, *i.e.* a study of how the 'intensity' of the particles (as measured by their rate of arrival at an appropriate detector) is a function of their energy. The most widely used of these is mass spectrometry, MS (see Box 5).

The combination of IR, NMR, and MS gives, almost always, enough information for the structure of a new or unknown compound to be worked out, and, with specialist instrument operators to help, a chemist who is proficient at interpreting the spectra will be able to sort out the structure of a molecule in a matter of hours. This is such a commonplace procedure these days that we should perhaps spare a thought for an earlier generation of chemists who established the structure of compounds by the painstaking stepwise synthesis of the molecule by an unambiguous route, with nothing more than simple glassware, pipettes, burettes, and balances to show them that they were on the right lines. Their practical technique must have been of the very highest order.

Some other particle spectrometries are described in Boxes 6 and 7. Although several of these techniques require special laboratory facilities, have a high capital cost, and may well require specialised knowledge to interpret the results, analytical chemists need to be aware of the capabilities of these techniques. There are several centres in the U.K. which 'hire out' these facilities, and thus they are available if the solution to the problem in hand requires the use of one of these types of spectrometry.

BOX 5   Principles of Mass Spectrometry

Mass spectrometry consists of breaking the sample into positively charged fragments, separating these fragments according to their mass : charge ratio (m/e), and measuring the relative abundances of the various m/e values. Like NMR spectroscopy, the technique is very powerful indeed at providing information about molecular structure; unlike NMR, it is very sensitive and can be used for trace analysis of both inorganic and organic species.

A large variety of methods are used for producing the ions from the sample, but the most common is to irradiate the sample vapour with a beam of electrons of sufficient energy to ionise the molecules and break the chemical bonds. There are several ways to separate the various m/e values, the most efficient being to pass the ions through a curved electrostatic field and then through a curved magnetic field. Only ions of the appropriate m/e value and velocity entering the system are deflected by the two fields to exactly the right extent to pass through the narrow slits at the beginning and end of the flight path. In order that collisions with neutral gas molecules do not deflect the ion beam to any extent, the entire instrument is operated under an ultra-high vacuum. This makes for additional complexity in the construction, operation, and maintenance of mass spectrometers.

As with NMR instruments, specialist operators are required. Interpreting the spectrum takes a bit of practice, but the basic idea is to 'reconstruct' the original molecule by piecing together the various fragments that are observed in the spectrum. Developments in this field of spectroscopy have been dramatic over the last few years, with the introduction of instruments with computer-based analysis (usually by comparing the spectrum obtained with spectra stored in the computer memory or accessed from some suitable data bank, maybe even via a satellite link) of the spectra obtained from complex mixtures after separation by a chromatographic technique (see Chapter 6). Analyses, *e.g.* for trace organic pollutants in water, that a few years ago would have been regarded as impossible are now routinely conducted.

Another exciting development is the use of an inductively coupled plasma (see page 39) as an ion source for a mass spectrometer, providing extremely low limits of detection with little mutual interference for a wide range of elements.

BOX 6   Electron Spectrometry

There are a number of analytical techniques based on the emission of electrons from a material. The most useful are those concerned with the examination of solid surfaces. Irradiating a solid with ionising radiation (such as X-rays or even an energetic electron beam) causes electrons to be removed from core and outer levels. These electrons can only escape from very shallow depths (a thickness of just one or two atoms), and so any electrons which are detected must have come from the top atomic layers of the surface. Just as with X-rays, the energies of the emitted electrons are characteristic of the particular element and their intensity is related to the relative amount in the surface. This technique is known as X-ray photoelectron spectrometry (XPS). XPS also provides chemical information because the exact energy of a core level is dependent on the charge distribution around the atom, which depends on the chemical environment and the bonding. Thus XPS spectra show chemical shifts, a bit like NMR chemical shifts, that can provide useful information. Another process which occurs as electrons from lower-energy orbitals fill vacancies in higher energy orbitals is the ejection of an electron, rather than of an X-ray photon. This process is known as the Auger (French pronunciation) process and provides similar information to that obtained from XPS, though with the possibility of greater spatial resolution. That is, it is possible to obtain 'maps' of the two-dimensional distribution of elements in a surface.

BOX 7   α- and β-Particle Emission

Radioactive nuclei decay by a variety of mechanisms. These include the emission of γ-rays (see page 48), helium nuclei (α-particles), and electrons or positrons (β-particles). Depending on the composition of the sample, all three types of emission may follow activation by neutrons. However, γ-ray spectrometry is the most used method of measurement in neutron activation analysis. Both γ-ray and β-particle emissions are used in a number of isotope dilution procedures. The basic principle behind these methods is to add a known amount of a radioactive version, which is of known activity, of the species being sought and allow complete mixing to occur. A known amount of the species is then isolated from the mixture and its activity measured. It is then a simple matter to calculate the amount of inactive species present in the original sample. This method depends on using another method to find out how much of the species has been isolated, but the isolation step (or steps) does not have to be 100% efficient and thus rapid (but inefficient) procedures can be used.

The same two disintegration products are used in radioimmunoassay procedures (see page 137). The basis of these methods is the specific reaction between some molecules and the antibodies raised to them in experimental animals. A radioactive version of the molecular species sought is produced by incorporating a radioactive atom into the molecule (*e.g.* ^3H, a β-emitter, or ^{125}I, a γ-emitter). These labelled molecules are added to the sample and compete with the unlabelled molecules of the species sought for the binding sites on an antibody reagent. After separation of the reacted molecules the activity is measured. For a given amount of labelled species the activity of the reaction product will decrease with increasing amount of unlabelled species.

THE END OF THE RAINBOW?

As you can see, the information that can be obtained by measuring the extent of the interaction of electromagnetic radiation with matter as a function of the energy (wavelength, frequency, or wavenumber) is considerable. Spectroscopic methods are undoubtedly one of the analytical chemist's most useful tools, and developments have been such that one might easily consider that we have reached the 'pot of gold' in terms of what spectroscopic methods can do. However, this is probably not so, as there is a considerable amount of research and development going on all over the world that is bound to bear fruit. Analytical chemists will be watching out for the applications of lasers and fibre optics and for the development of body scanners, organ imagers, and other non-invasive techniques to aid medical diagnoses.

Many analytical chemists are striving to improve the limits of detection of some of these techniques. There is still a long way to go before techniques capable of detecting single chemical entities are in routine use in analytical laboratories.

FURTHER READING

J.S. Fritz and G.H. Schenk, 'Quantitative Analytical Chemistry', 5th Edn., Allyn and Bacon, Boston, 1987.

H.H. Willard, L.L. Merritt, jun., J.A. Dean, and F.A. Settle, jun., 'Instrumental Methods of Analysis', 6th Edn., Wadsworth, Belmont, California, 1981.

G.T. Bender, 'Principles of Chemical Instrumentation', W.B. Saunders, Philadelphia, 1987.

'Introduction to Ultraviolet and Visible Spectrometry', 2nd Edn., ed. J.E. Steward, Pye Unicam, Cambridge, 1985.

'Standards in Absorption Spectrometry', ed. C. Burgess and A. Knowles, Chapman and Hall, London, 1981.

'Practical Absorption Spectrometry', ed. A. Knowles and C. Burgess, Chapman and Hall, London, 1984.

R.C.J. Osland, 'Principles and Practices of Infrared Spectroscopy', 2nd Edn., Pye Unicam, Cambridge, 1985.

R.J. Taylor, 'The Physics of Chemical Structure', Unilever, Blackfriars, 1971.

D.H. Williams and I. Fleming, 'Spectroscopic Methods in Organic Chemistry', 3rd Edn., McGraw-Hill Book Company (U.K.), London, 1980.

L. Ebdon, 'An Introduction to Atomic Absorption Spectroscopy', Heyden, London, 1982.

'Standards in Fluorescence Spectrometry', ed. J.N. Miller, Chapman and Hall, London, 1981.

D.T. Burns, A. Townshend, and A.G. Catchpole, 'Inorganic Reaction Chemistry; Systematic Chemical Separation', Ellis Horwood, Chichester, 1980.

Chapter 3

Making Electrons Work

In the previous chapter the many ways in which the interaction of various forms of electromagnetic radiation with atoms and molecules can provide analytical information were described. In all these techniques an instrument was used that at some stage converted the light intensity into an electrical quantity, which was then processed by the appropriate circuitry before the final readout was displayed. The conversion from light energy to electrical energy is inefficient and so, in principle, it should be possible to achieve higher sensitivities and lower limits of detection by using the electrical properties of the sample directly. However, despite the fact that very small values of charge, current, potential, and capacitance can be measured quite reliably, electrochemical methods of analysis have not achieved the kind of widespread use that spectroscopic methods have, apart from one or two specialised methods which will be described shortly.

If an electrical method is to give some analytical (or any other) information about a sample, then the sample components must have some electrochemical properties, *i.e.* they must be electroactive. The components must either lose or gain electrons fairly readily so that, in solution, the species are ionized (either positively or negatively charged) or can accept or donate electrons at electrode surfaces (or both) in order to interact with an external measuring electrical circuit. Very little information can be obtained from the electrical properties of a solid sample, so all electroanalytical methods involve a solution (usually in water) of the sample.

Contact between the solution and the external electrical circuit is obviously necessary, and this is done with electrodes (at least two, of course). There are a large number of different sorts of electrodes, and it is really the design of these that makes electroanalysis possible. The analytical chemist must understand how these various types of electrodes can be used as the basis of analytical methods and what information may be obtained from them. This, together with an understanding of the limitations of each electrode's performance, will allow the analytical chemist to make a sensible choice from among the available methods for the problem being tackled, always assuming that electroanalytical methods are applicable at all.

It is important that the sample solution is electrically conducting. Not a lot of useful analytical information is obtained from measurements on solutions which, together with the electrodes, act as a capacitor in the circuit. An electrically conducting solution contains ions, and if the sample does not provide sufficient of them they may have to be added. Such an additional electrolyte is known as a background electrolyte. Potassium sulphate or chloride is often used for this purpose.

RECIPROCAL RESISTANCE

Although it may appear that the resistance of a sample solution is a fairly fundamental property, it has only limited use. Electrochemists prefer to deal with the reciprocal of resistance, which is called conductance. Its measurement is routinely used for checking for the absence of ions, for example in monitoring the quality of high-purity water for use in power station boilers or in analytical laboratories to make solutions for trace analyses. Unless special precautions are observed, the conductance of laboratory water, of any sort, will gradually increase as certain amounts of carbonate and hydrogen carbonate are produced by the reaction of atmospheric carbon dioxide with the water. This makes the water slightly acid. Sulphur dioxide and oxides of nitrogen will also dissolve and undergo hydrolysis. These gas–water reactions are partially responsible for the acidity of 'acid rain'. The water will also contain dissolved oxygen (which does not react with the water) at a concentration of about 10^{-3} mol dm^{-3}. This can prove a nuisance in some circumstances as oxygen is fairly easily reduced at an electrode to give hydrogen peroxide, which may be further reduced to water. On the other hand, these electrode reactions form the basis of a method for measuring dissolved oxygen, a vital parameter (quite literally!) in the monitoring of the quality of surface water.

Conductance measurements are non-specific, *i.e.* they only provide information about the total concentration of ions in the solution. Such measurements can be used to follow the course of a reaction in which there is a change in conductance due to a change in the number of ions or in the mobility of the ions. Ions which can move quickly, such as H^+ and OH^-, provide a higher conductance than the same concentration of slower-moving ions. For example, the amount of a metal ethanoate in a solution could be determined by following the change in conductance as hydrochloric acid is added. The plot is shown in Figure 3.1. As the reaction proceeds, the conductance increases slightly because CH_3COO^- is effectively replaced by Cl^- as the ethanoate is converted into undissociated ethanoic acid. Once this is complete, the conductance increases sharply as the concentrations of H^+ and Cl^- increase.

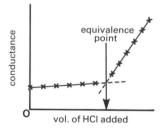

Figure 3.1   *Determination of the amount of a metal ethanoate by following its reaction with hydrochloric acid. An equivalent amount of HCl has been added when the discontinuity in the plot occurs. To locate the equivalence point an excess of HCl is added and the plot is extrapolated back.*

Conductance measurements are also used to measure the amounts of ions in mixtures after they have been separated. For example, the separation of inorganic anions by a chromatographic method (see Chapter 6) followed by conductimetric detection has proved a very successful method of analysing such mixtures, and it is possible to buy chromatographs designed specifically for this purpose.

METHODS WITH POTENTIAL

Whenever a solid is placed into a solution of ions, there will be a potential difference between the solid and the solution. The analytically useful situations are those in which oxidation or reduction reactions involving the transfer of electrons can take place at the solid surface. Analytical methods which use the electrochemistry of such redox reactions are known as potentiometric methods. This classification also includes those methods in which a potential difference between a surface and solution arises because of migration of ions from the surface into the solution (or the other way round). The problem is that this potential difference cannot be measured without the insertion of another solid into the solution, and unless this electrode is different from the first there will be no potential difference between them anyway. If there is a potential difference between the two electrodes, it will only be measurable if the electrodes are conducting (even if only to a very small extent). Any potential difference will not last long with this set-up, as reactions between the solution and the electrodes will occur to make the potential differences at each electrode the same. If we are going to measure the potential difference between our first solid and the solution we must make sure that our second solid does not come into contact with the solution.

Sounds impossible, doesn't it? However, the important thing is to avoid contact between certain ions in the solution and the solid, and it *is* possible to do *this*. We put the second electrode into a second solution and connect the two solutions by a device called a salt bridge. Salt bridges are made of a variety of materials, a common one consisting of a concentrated solution of potassium chloride in a gel made from agar (a complex carbohydrate obtained from certain marine algae) of about the consistency of thick wallpaper paste. The bridge allows electrical contact between the two solutions by means of K^+ and Cl^- ions moving across the bridge but also provides an effective barrier to any other ions. The situation is shown in Figure 3.2. This entire system is often known as an 'electrochemical cell' or just 'cell' for short.

We now have a reading on our voltage measuring device, but we are not much nearer to the object of the exercise (which was to measure the potential difference between solid 1 and solution 1,

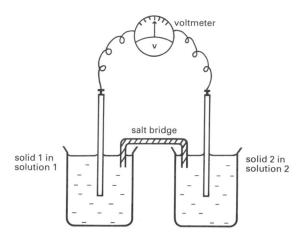

Figure 3.2 *Measuring the potential difference between electrodes and solutions.*

which we expect to be a function of the concentration of the ions in solution 1 and therefore to be the basis of an analytical method). This is because the potential difference measured on the voltmeter is the net effect of the potential difference between solid 1 and solution 1 and between solid 2 and solution 2. There is no way around this difficulty: all electrode potentials (*i.e.* the difference in potential between the solid and the solution in contact with it) can only be measured with reference to another electrode. The term 'electrode' is used rather loosely by electrochemists. It can mean simply a metal (like copper or platinum) or it can mean a metal in contact with a solution, or a metal in contact with an insoluble salt in contact with a solution, or even a metal in contact with an insoluble salt in contact with a solution in contact with a membrane! And there are arrangements even more complicated than this which are referred to as 'electrodes'.

In order to assign numbers to single-electrode potentials, a standard value has to be agreed. Just as tetramethyl silane was assigned a chemical shift value of zero in NMR spectrometry (see page 46), the agreed reference electrode is assigned a value of zero volts. The particular electrode chosen as the zero point is the electrode in which hydrogen gas at a pressure of one atmosphere is in contact with a 1.228M solution (at 20 °C) of hydrogen ions (provided by HCl). This contact is effected by bubbling the gas over a platinum electrode coated with platinum black (finely divided platinum). This standard hydrogen electrode (SHE) is not exactly the most convenient electrode to use, so in making measurements in the laboratory other reference electrode–solution combinations are used. The most commonly used such electrode is the saturated calomel (Hg_2Cl_2) electrode (SCE). This is shown in Figure 3.3. The SCE has a potential of -0.242 V with respect to the SHE. This potential depends on the amount of chloride in the solution, which will vary with temperature, so the potential is a function of temperature.

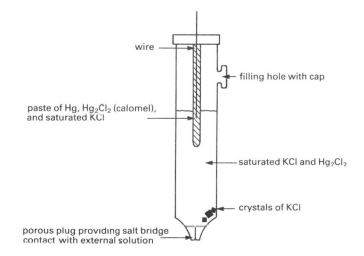

Figure 3.3 *The saturated calomel reference electrode.*

In general, electrode potentials of analytically useful electrodes are related to concentration by a relationship known as the Nernst equation:

$$E = E^\circ \pm \frac{RT}{nF}\ln[\text{X}] \tag{3.1}$$

where E° is the potential (in volts) under standard conditions of temperature and concentration, R is the gas constant, T is the absolute temperature, n is the charge on the ion, and [X] is the concentration of the species being analysed. If X is a cation, the '+' sign applies; if X is an anion, the '−' sign applies. Converting to logarithms to base 10 and inserting values for R, T, and F gives at 25 °C:

$$E = E^{\oplus} \pm \frac{0.059}{n} \log[X] \qquad (3.2)$$

In making an analytical measurement we are measuring a cell potential (*i.e.* the potential difference between two electrodes). One of the cell electrodes will be the ion-sensitive electrode and the other will be the ion-insensitive electrode (usually because it is a reference electrode connected to the solution by a salt bridge). The cell potential will be made up of the two electrode potentials [each of which will have its own version of equation (3.2)] together with any potentials arising because of the different rates of movement of ions across a boundary and any potential arising due to the resistance of the circuit (from Ohm's law a current i flows in a resistance R because the driving force is a potential difference equal to iR).

The usual method of using such cells for analytical measurements is to keep everything in the cell constant apart from the concentration of the ion to be measured. So the expectation is that from equation (3.2) a factor of ten change in concentration of the ion would result in a change in potential of $0.059/n$ volts. That is, for a singly charged ion the change would be only 59 mV.

Ion-sensitive electrodes can be used for following the course of a titration and indicating the end-point, around which large changes in concentrations occur (see Chapter 4). For making direct measurements the electrode has to be calibrated. That is, the potential is measured for a number of solutions of known ionic concentration and a graph of potential against log(concentration) is plotted. The concentration of the unknown solution is found by interpolation. Most electrodes in routine laboratory use require frequent recalibration.

One of the most widely used types of electrode is the membrane type shown in Figure 3.4. The circuit will consist of another electrode (a reference electrode) and a potential measuring device. This will normally be a voltmeter of some sort, which draws very little current indeed. This is required because the resistance of the membrane may be very high and thus the 'iR' potential could swamp the ion exchange potential. The purpose of the internal reference electrode and solution is to maintain electrical contact at a fixed potential with the internal side of the membrane. With some membranes the wire can simply be glued to the back of the membrane.

Even though the membrane produces a potential because of an

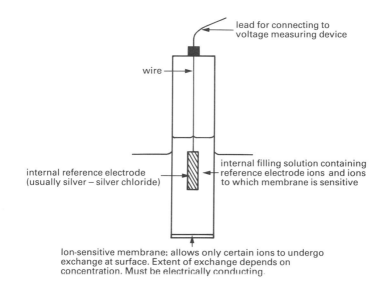

lead for connecting to voltage measuring device

wire

internal reference electrode (usually silver – silver chloride)

internal filling solution containing reference electrode ions and ions to which membrane is sensitive

Ion-sensitive membrane: allows only certain ions to undergo exchange at surface. Extent of exchange depends on concentration. Must be electrically conducting.

Figure 3.4 *A membrane-type ion-sensitive electrode. If the membrane only responds to certain ions, the electrode is known as an ion-selective electrode.*

ion exchange mechanism not a redox reaction, the Nernst equation still holds and a $59/n$ mV change is expected for every ten-fold change in concentration. By far the most widely used type of membrane electrode is the glass electrode, whose membrane consists of a special glass (63% SiO_2, 28% Li_2O, 5% BaO, 2% La_2O_3, and 2% Cs_2O) which is highly selective for hydrogen ions once the glass has been soaked and acquired a hydrated gel layer on the surface acting as an ion exchanger for H^+. Normally the membrane is spherical, as the easiest method of construction is to 'blow' the ion-sensitive glass on the end of an ion-insensitive insulating glass tube. Quite often the pH electrode (as they are often called) body will contain the second reference electrode needed to complete the circuit, so to make a measurement only one device needs to be inserted into the solution.

There are a large number of other ion-sensitive electrodes (ISEs) commercially available for a whole range of commonly encountered inorganic cations and ions. Apart from the fluoride electrode, the performance of the others is a long way behind that of the pH electrode. Unlike most other analytical techniques,

ISEs have a very wide calibration range, *i.e.* the concentration ratio of the top standard to the bottom one may be 10^5 or more (the range for the pH electrode can be 10^{13}). Most spectroscopic methods such as solution spectrometry or atomic absorption spectrometry can only cope with a ratio of 100 or so. This wide 'dynamic range', as this feature is called, is due to the logarithmic response of the electrode. The response usually extends to the trace level, maybe 10^{-6} mol dm^{-3} or less.

Equation (3.2) is strictly only valid for the ion 'activity' (see page 80) rather than concentration. This can be useful, for example in clinical analysis where the biological function of the ion is more closely controlled by its 'activity' than by its total concentration (which is what might be measured by another analytical technique). However, a small error or fluctuation in the potential means quite a large uncertainty in the concentration (or activity) measured.

Most ISEs are not very selective and respond to other ions as well as the one they were designed for. This can be a severe problem if the interfering ions are present in high concentration. Nor do the electrodes last very long, maybe only a few weeks or months. A fair amount of research is being devoted to developing improved ISEs, but the results so far have not been particularly encouraging, so this field offers quite a challenge to polymer and organic chemists as it's the design of the membrane and ion-selective material that's important. It is possible to make very inexpensive ISEs by simply coating a copper wire with the ion-sensitive membrane. This 'throw-away' approach may be more promising than trying to make electrodes with lifetimes and performances comparable with those of the glass electrode.

CURRENT DEVELOPMENTS

The type of electrochemical cell that is used for the basis of potentiometric measurements is called a 'galvanic' cell. In effect the cell is a source of electrical energy because the electrode processes are those which occur spontaneously. If an external voltage source is used to drive the current round the circuit and force particular reactions to occur at the solid–solution interfaces, then the cell is known as an electrolytic cell.

In some special cases it is possible to determine how much of a particular chemical is in the cell by measuring the amount of electricity required for it to be completely electrolysed. The basic quantitative relationship is known as Faraday's law of electrolysis, which states that:

$$w = M_r it/nF \qquad (3.3)$$

where w is the mass (g) of material resulting from the electrolysis, M_r is the relative molecular mass of the ion involved, i is the current (amps), t the time (s), n the number of electrons per ion, and F is Faraday's constant. If the experiment is not done at constant current, the total quantity of electricity in coulombs has to be substituted for it. Such 'coulometric' methods are not widely used. In fact, no method based on electrolytic cells is particularly widely used, but those that are used are based on measuring current–voltage relationships.

These methods are known as 'voltammetric' (because the voltage between the electrodes is scanned and the current measured) and can use a variety of electrodes. From the analytical point of view the most information is obtained from a rather special electrode, the dropping mercury electrode (DME). The DME is just that: a reservoir of mercury (connected to the external circuit) forces drops to continuously form and fall off the tip of a fine capillary tube dipping into the solution to be analysed. Usually the circuit is completed using two other electrodes. One is a reference electrode (SCE, say) and the other a piece of platinum. A reference electrode is needed so that the potential of the working electrode can be controlled properly, but it's not a good idea to draw current through a reference electrode as this alters its potential, so all the current passes through the DME and the third electrode, which is known as the counter-electrode. The circuit is shown in Figure 3.5.

The technique is known as polarography and the device a polarograph. The resulting plots of current against voltage are polarograms. What do they look like? The current we want to measure flows because of reduction of electroactive species in solution at the DME. Because the DME is small compared with the volume of the bulk solution it is possible to produce a situation in which a change in potential no longer produces an increase in current. This is because the current is controlled by the rate of arrival of the electroactive species at the electrode surface by the process of diffusion. Once the current that corresponds to the

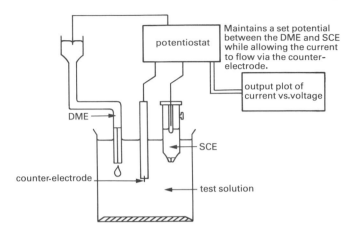

Figure 3.5 *Basic circuit for a three-electrode polarograph. Dissolved oxygen must be removed by bubbling out with nitrogen, and the solution must contain about 0.1M KCl or similar to provide adequate conductance. The solution must be neither stirred nor shaken during electrolysis.*

fastest diffusion rate is being passed, no further increase can be obtained. Superimposed on this are the fairly rapid oscillations in current that occur as the drops grow in size and fall off. Part of the current which flows is required to charge the drops, as the cell also has a certain capacitance. The situation is summarised in Figure 3.6.

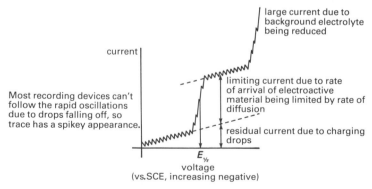

Figure 3.6 *A typical d.c. polarogram. The DME is normally scanned in a negative direction versus the SCE as mercury dissolves (with a very high current flow) at positive potentials. The potential half-way up the rising part of the wave is called the half-wave potential, $E_{1/2}$.*

The diffusion-limited current, i_d, is directly proportional to the concentration, c, of the electroactive material in the solution. The full equation (known as the Ilkóvič – pronounced 'Ilkovitch' – equation) involves a number of physical constants of the system, including the rate of mercury flow, but can be summarised as:

$$i_d = kc \tag{3.4}$$

where k is a constant for a given set of conditions. Most of the reactions are reductions, in which the DME acts as a source of electrons. There is no interference from build-up of deposited material at the electrode because the surface is continually renewed as the drops fall off. This is what makes polarography a useful technique. Most voltammetric techniques using solid electrodes eventually grind to a halt because the working electrode surface has become contaminated. At this stage the working electrode has to be cleaned. Mercury also resists the evolution of hydrogen gas at its surface (the reaction has a high activation energy – see page 78) so that potentials may be set at which it would be expected that the major current-carrying reaction would be $H_3O^+ + e^- \rightleftharpoons \frac{1}{2}H_2 + H_2O$. Although mercury is difficult to handle and is toxic, no-one has yet found a better electrode material in the fifty years since polarography was invented.

Although the $E_{1/2}$ value is characteristic of the particular ionic species being reduced, polarography is not a serious contender as a method of qualitative analysis. It does mean, though, that, if the solution contains species with sufficiently different $E_{1/2}$ values, they can be determined by measuring the limiting current for each of the waves. As with other techniques, polarographs have to be calibrated. This is most conveniently done by the standard additions method (see page 132) in which, after running the polarogram for the sample, a small amount of a concentrated standard solution is added and the polarogram re-run.

The basic d.c. version of the technique has a somewhat limited performance as a trace technique because of the fairly large residual or charging current. The effect of this can be decreased by applying a potential–time relationship different from the simple linear ramp used in the d.c. mode. The most successful of these variants is differential pulse polarography, DPP (see Box 8), whose limit of detection may be considerably better than d.c. polarography (maybe a factor of 100 or more), and in favourable

BOX 8   Differential Pulse Polarography

The basis of DPP is shown in Figure 3.7. Small pulses are superimposed on the basic ramp and synchronised with the end of the drop lifetime (actually it's the other way round – the drops are synchronised with the pulses by mechanically knocking them off every few seconds), and the current is measured just before the pulse and just at the end of the pulse. The difference between these two currents is plotted as a function of voltage. The idea is to measure the current near the end of the drop lifetime, where the rate of change of surface area is least (the charging current grows as the electrode area grows), and at the end of the pulse, when the additional charging current due to the small rapid increase in potential has decayed but the current due to electrolysis is still increasing. The charging current is to a large extent subtracted out and the difference corresponds mostly to electrolysis current.

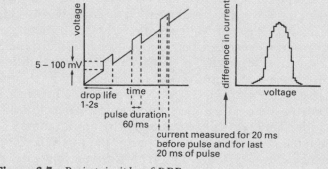

Figure 3.7 *Basic principles of DPP.*

cases the detection limit can be as low as 10^{-8} mol dm^{-3} of the electroactive species.

DPP has been used for the determination of certain metals at the µg dm^{-3} level in natural waters, but the preferred method for this sort of analysis is atomic absorption spectrometry with a carbon furnace atomisation device (see page 36). It is possible to determine organic compounds by polarography if they contain a reducible function group, and DPP has been applied to the study of the degradation of synthetic food colours, several of which contain a reducible azo (N=N) group. Polarographic techniques are most likely to find application in organic rather than inorganic analyses.

There is a third type of method based on the measurement of current that has a lower detection limit than either coulometry or any of the variations on the polarography theme. This is called

BOX 9 Stripping Voltammetry

Two types of stripping voltammetry (SV) are known, anodic SV (ASV) and cathodic SV. ASV, the more frequently used, has two distinct stages. Firstly, the reducible species in the solution (usually a metal ion) is electrodeposited onto a small electrode (quite often just a hanging mercury drop). This process can go on for some time, and, if there is a lot of solution for analysis, even very low concentrations will eventually lead to a build-up of electrodeposited material at the electrode. Many metals form amalgams (alloys with mercury), so they dissolve and diffuse into the drop and leave the surface practically unaltered. Secondly, the potential is scanned in order to dissolve the metals out of the mercury drop back into the solution (sometimes the solution in the electrolysis cell is changed prior to the scan). The metals are being oxidised, so the electrode is acting as the anode (whereas it was the cathode in the first part), and stripped out of the drop into the solution. The current is measured as a function of the stripping potential. Hence the term anodic stripping voltammetry. The detection limits can be at the sub μg dm^{-3} level, and the technique rivals atomic absorption spectrometry using electrothermal atomisation (see page 36) in this respect. ASV only applies to metals which can be reduced at a mercury electrode, but quite a few of these are the ones that we want to keep an eye on at such low levels, such as lead and cadmium. An example of an ASV trace is shown in Figure 3.8. The peak height and the area under the peak are both proportional to the amount of metal deposited.

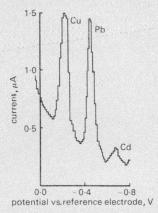

Figure 3.8 *ASV measurement of cadmium, lead, and copper in a Baltic sea-water sample. The deposition time was 10 min. The cadmium and lead concentrations were found to be 73 and 980 ng dm^{-3}, respectively.*

(Reproduced with permission from I. Gustavsson and K. Lundström, *Talanta*, 1983, **30**, 959.)

stripping voltammetry, SV (see Box 9). There are a number of variations on the SV theme which are currently (!) being developed, the main interest being the very low detection levels that can be achieved.

FURTHER READING

J. Vesely, D. Weiss, and K. Stulik, 'Analysis with Ion-Selective Electrodes', Ellis Horwood, Chichester, 1978.

A.M. Bond, 'Modern Polarographic Methods in Analytical Chemistry', Marcel Dekker, New York, 1980.

F. Vydra, K. Stulik, and E. Julakova, 'Electrochemical Stripping Analysis', Ellis Horwood, Chichester, 1976.

Chapter 4

Analytical Reaction Chemistry

In Chapters 2 and 3 the use of chemical instruments by the analytical chemist for making spectroscopic or electrochemical measurements was discussed. Both of these areas of analytical chemistry are applied for the determination of trace components of materials.

However, trace components are not the only ones about which analytical chemists are asked to provide information. There are many situations in which the analysis of the minor (0.1—1%) or major (1—100%) components is required. In general, different techniques are used for determining major components; the same techniques may be used for the determination of minor components or the techniques of Chapters 2 and 3 may be applied. When several techniques could possibly be used, the analytical chemist has to choose the most appropriate one for the job in hand (see Chapter 7).

When the job in hand is measuring how much copper in brass, or lead in solder, or protein in a meat pie, or detergent in a washing powder, or aspirin in a pain-relieving tablet, the analytical chemist is usually asked to provide accurate information with as little uncertainty about it as possible. This means, for example, that a method for determining copper in brass which involves dilution of a sample solution by factors of several hundred (in order to measure the copper by atomic absorption spectrometry) will not be satisfactory because of the uncertainties introduced by the graduated glassware (see page 120).

The analytical chemist must therefore make use of methods which are accurate and have a low level of uncertainty associated with them for the determination of major components. There are two physical properties which have these desired characteristics and which can be measured readily in the analytical laboratory. Thus the determination of major components is performed by analytical techniques based on measurements of volume or mass.

It is possible to determine the volume of liquid delivered from a burette to the nearest 0.01 cm^3 and to weigh an object on a laboratory balance to the nearest 0.1 mg. For a volume of 50.00 cm^3 and a mass of 100 g this represents uncertainties of 1 part in 5000 and 1 part in 1 000 000, respectively. Analytical chemists therefore make use of chemical reactions which allow quantification of sample composition to be made by the measurement of volume or mass. The analytical methods based on such reactions are known as titrimetry (or volumetric analysis) and gravimetry, respectively.

REACTIONS FOR TITRIMETRY AND GRAVIMETRY

In principle, reactions for titrimetry and gravimetry are similar to those employed as the basis of a spectroscopic method, such as the formation of a coloured product for the determination of orthophosphate by solution spectrometry by reaction with molybdate(VI) ions in acid solution followed by addition of a reducing agent [such as tin(II) ions]. In general, we can refer to such reactions as the addition of a reagent, R, to a solution of the species to be measured, X, to form a product, XR:

$$X + R \rightleftharpoons XR \qquad (4.1)$$

The concentration of XR will be equivalent to the initial concentration of X, provided enough R is added. A large excess of R may be used, provided it does not interfere with the measurement of XR, and thus the equilibrium constant for reaction (4.1) does not have to be very high. The same could be said if the mass of XR formed was to be used as the basis of the analysis. All that is required is that sufficient R is added so that essentially all of X is converted to XR.

However, this is not the situation required when the volume of the solution containing R is the basis of the quantitative measurement, as here excess R cannot be added, otherwise a gross overestimation of the amount of X will be made. For titrimetric analysis the equilibrium constant of the reaction must be very large so that the first excess of R added after an amount equivalent to X has been added may be detected to indicate that the equivalence condition has been achieved.

In addition to the constraints on the value of the equilibrium constant of a reaction for an analytical purpose there are two other constraints that analytical chemists must consider. Firstly, the speed of reaction is important. Generally the faster it is, the better, and again the constraints of titrimetric analyses are probably the most stringent. The reaction should proceed at least as fast as R is added to X. Secondly, the selectivity of a reaction must be considered, as rarely does the analytical chemist face the situation in which the sample solution contains only one species. In most cases the solution will contain other potentially interfering species, and the analytical chemist must know to what extent the primary reaction is going to be subject to unwanted side reactions. Some examples of these were discussed in connection with the detection of lead in river water (see page 21).

In devising an analytical method for a particular problem, analytical chemists are not often involved with finding a new primary reaction to use; what they are often involved with is designing a set of reaction conditions to make a standard primary reaction specific in the particular mixture being studied. Analytical chemists also need to know the limitations of the primary reactions under the conditions employed for the particular analysis, so that a reliable statement about the uncertainty of the result obtained may be made.

Thus analytical chemists need to understand the factors which control (*a*) the extent to which a reaction occurs and (*b*) the speed of a reaction.

Theories Concerning Chemical Reactions

Analytical chemists, just as any other sorts of chemist, need to have an answer to the question 'Why do chemical reactions occur?' Although the question appears simple, the answer is one of the central themes of chemistry which has fascinated and exercised the minds of several generations of chemists.

To simplify the discussion here somewhat, we will confine our attention to reactions between species in aqueous solution. These are not the only reactions that analytical chemists make use of, but the concepts involved for such reactions are applicable to other sorts of reactions (such as solid–solid, solid–gas, and gas–gas reactions).

First of all, we need a model for a chemical entity in water. We can think of water molecules moving randomly within the vessel with a spread of velocities. The average speed and shape of the speed distribution curve depend on temperature. The higher the temperature, the faster the molecules move. Water does not have long-range structure, such as a crystalline solid, nor are the molecules confined to moving in a small volume, as in a solid, but there are attractive forces between molecules sufficient to prevent water at room temperature being a gas.

Introducing chemical entities into water involves the expenditure of energy to make the appropriately sized holes in the water structure. For species which are considered soluble, this is more than offset by the energy released due to the strong bonds which form between a number of water molecules and the solute entity (the hydration energy). The model is summarised in Figure 4.1.

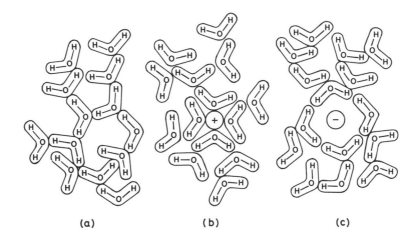

(a)   (b)   (c)

Figure 4.1   *Models for water. (a) Pure water. H_2O molecules contain positive charges on the hydrogens and negative charges on the oxygens. The molecule has a sort of banana shape because of the repulsion between the electrons forming the O–H bonds and the non-bonding electrons localised on the oxygen. (b) Positively and negatively charged ions in aqueous solution. The ions orientate the water molecules because of the electrostatic interaction.*

At this stage let us assume that we have one solution containing some aluminium ions and another solution containing the fully ionised anion of 1,2-bis[bis(carboxymethyl)amino]ethane (edta^{4-}), which we mix together. Obviously there will be other ions present: the counter-ion associated with Al^{3+} and the H$^+$ from H$_4$edta. Edta is a hexadentate ligand forming 1 : 1 complexes with a range of metal ions (see Figure 4.2), and its use in analytical chemistry is discussed later in this chapter.

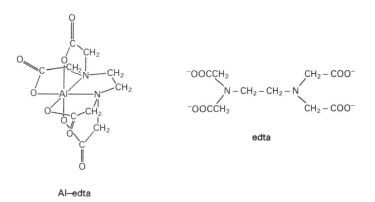

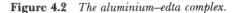

Al–edta

edta

Figure 4.2 *The aluminium–edta complex.*

The hydrated Al^{3+} and edta^{4-} will be buffeted by the water molecules, and when two such ions collide with sufficient energy to break the ion–water bonds they react and a product ion, Aledta$^-$, forms. This process occurs at a rate governed among other things by the concentrations of the ions. The product ions will also be in collision with species in solution, and when a sufficiently energetic collision with Aledta$^-$ occurs the starting materials, Al^{3+} and edta^{4-}, are reformed. Sooner or later the rate of formation of the product equals the rate of reformation of reactants and, as observers perceiving only the overall situation, we say that the reaction has reached equilibrium.

It is important to note that the reaction hasn't stopped (though quite often we use this term for a reaction which has reached equilibrium), it is still going on – it is just that the rate of the forward and that of the backward processes are equal.

If we consider the probabilities of the collisions leading to a change (both forward and backward), then we deduce a simple relationship between the equilibrium concentrations of the reacting species. In general, for the reaction:

$$aA + bB \rightleftharpoons cC + dD \qquad (4.2)$$

then:

$$\frac{[C]^c [D]^d}{[A]^a [B]^b} = K_c \qquad (4.3)$$

where K_c is the equilibrium constant and the square brackets denote equilibrium concentrations. As the rates of the forward and backward reactions are also a function of temperature, so K_c is also a function of temperature. For our particular reaction:

$$Al^{3+} (aq) + edta^{4-} (aq) \rightleftharpoons Aledta^- (aq) \qquad (4.4)$$

the equilibrium constant is given by:

$$K_c = \frac{[Aledta^-]}{[Al^{3+}] [edta^{4-}]} \qquad (4.5)$$

Large numbers of K_c values have been measured, and when we look up the value for K_c for reaction (4.4) we find the value 10^{17}. Thus at equilibrium the concentration of either Al^{3+} or $edta^{4-}$ (or both) is very small compared with the equilibrium concentration of $Aledta^-$, and in principle the reaction could be used as the basis of the titrimetric determination of Al^{3+} by $edta^{4-}$. The role of the value of K_c will be discussed later in this chapter.

Chemical reactions are just one example of a change in the physical world. Scientists of all disciplines are interested in changes, and a very large number of observations about all aspects of change have been made. In particular, scientists are interested in the direction of spontaneous change, with the position of equilibrium, with the energy gain or loss, and with the temperature dependence of these. The observed phenomena can be summarised in three statements, known as the three laws of thermodynamics (the science of energy transformations).

Chemical Thermodynamics. Although chemists are interested in all three laws, analytical chemists only need concern themselves with the first two: (i) energy can be neither created nor destroyed and (ii) the entropy of the Universe increases in the course of every spontaneous change. Entropy is a measure of the random dispersal of energy.

Chemists need to make use of the second law for two reasons. Firstly, it provides a 'motive' or driving force for a spontaneous change, namely that spontaneous change will occur as long as the entropy of the Universe is increased, *i.e.* as long as random dispersal of energy increases. We can combine this with the first law and say that we expect spontaneous chemical reaction to be accompanied by the production of heat. This liberation of energy warms up the solution, the containing vessel, the bench, and the air in contact with it, thus increasing the random thermal motion of all the species involved. The excess energy gradually leaks away and is randomly dispersed to increase the entropy of the Universe. We interpret the production of heat as being due to the fact that the chemical bonds in the product species are stronger than those in the reactant species. This is normally expressed as the energy change corrected for the result of any volume change. The reason for this is that a large number of chemical reactions are carried out at constant pressure, *i.e.* in vessels open to the laboratory atmosphere, and that if the reaction is accompanied by an increase in volume, say, then energy has to be expended in pushing back the surroundings, and this has to come from somewhere. This corrected energy change is referred to as the enthalpy change and given the symbol ΔH.

The second reason why chemists need to make use of the second law of thermodynamics is that not every reaction which proceeds spontaneously is exothermic (*i.e.* gives out heat, for which $\Delta H < 0$). There are some which are endothermic: the $Al^{3+}/edta^{4-}$ reaction is a good example, being endothermic to the tune of about 50 kJ mol^{-1}. In other words the aluminium–water bonds are much stronger than the aluminium–edta bonds. How does the second law account for a spontaneous endothermic reaction?

We need to look more closely at our 'universe' and in particular at the water molecules. Both the Al^{3+} and edta^{4-}, the reactants, are hydrated. In particular the Al^{3+} strongly binds several water molecules in a primary solvation layer and induces order in

several molecules in a secondary layer; thus a hydrated aluminium ion behaves as a single entity. When the product forms,
although it will be hydrated to some extent, the net effect is a
considerable increase in the number of randomly moving
solvent-type water molecules. Thus the 'universe' associated with
the product has a greater ability to disperse energy randomly (*i.e.*
greater entropy) than the 'universe' associated with the reactants.
Therefore, although the amount of energy associated with the
product state is less than that associated with the reactant state,
this is more than offset by the product 'universe's' greater
entropy, and thus an endothermic reaction proceeds spontaneously.

Thermodynamics summarises the overall situation by means of
equation (4.6):

$$\Delta G = \Delta H - T\Delta S \qquad (4.6)$$

where ΔG is known as the Gibbs function (or Gibbs free energy), T
is temperature, and ΔS is the entropy change for the reaction. Just
as for values of ΔH, there are compilations of ΔS values, usually
accompanied by the results of the calculation of ΔG. Tabulated
values are often calculated with reference to standard states and
are denoted by a superscript $^\circ$ following the appropriate symbol.

Spontaneous change occurs if $\Delta G < 0$. At equilibrium $\Delta G = 0$,
i.e. there is no longer any net driving force and no further overall
change occurs. The second law of thermodynamics is discussed
further in Box 10.

We now have a way of deciding whether any reaction we care to
write down could possibly be the basis of a titrimetric method of
analysis. All you have to do is locate values of ΔH° and ΔS° in
compilations of data, calculate ΔG°, and hence K_c (see Box 10).
Naturally analytical chemists will be capable of performing such
calculations, but they will also be capable of interpreting the
values of ΔH° and ΔS° they have calculated for a particular
reaction in terms of what is actually going on in the reaction
vessel.

Fast Reactions

The analytical chemist who now sets out to measure the amount
of aluminium in a sample of bauxite (an ore whose composition
approximates to $Al_2O_3 \cdot xH_2O$) by titration of a sample solution

BOX 10   The Second Law of Thermodynamics

The second law states that, for a spontaneous change to occur, the entropy of the universe must increase.

Entropy is defined by equation (4.7):

$$\Delta S = q/T \tag{4.7}$$

where ΔS is the entropy change caused by the transfer of an amount of heat energy q, reversibly at temperature T. Unlike enthalpy changes involved in chemical reactions which mainly concern the energies of bonds broken and made in the reaction vessel, entropy changes affect both the surroundings and the contents of the reaction vessel. The change in entropy of the surroundings, ΔS_{surr}, is simply related to the enthalpy change, ΔH, by equation (4.8):

$$\Delta S_{surr} = -\Delta H/T \tag{4.8}$$

Thus, if the reaction is exothermic, for which ΔH is negative, there is an increase in entropy of the surroundings. The total change in entropy, ΔS_{tot}, is simply the sum of the change in entropy of the surroundings and the change in entropy of the reaction, ΔS:

$$\Delta S_{tot} = \Delta S_{surr} + \Delta S \tag{4.9}$$

Substituting from equation (4.8) for ΔS_{surr} and rearranging gives·

$$-T\Delta S_{tot} = \Delta H - T\Delta S \tag{4.10}$$

Thus, for a chemical reaction to occur spontaneously, ΔS_{tot} must be positive, and hence from equation (4.10) $\Delta H - T\Delta S$ must be negative.

The function $-T\Delta S_{tot}$ is referred to as the Gibbs free energy (or Gibbs function) and is given the symbol ΔG. Thus the equation from which it is possible to deduce the direction of spontaneous change of a reaction is $\Delta G = \Delta H - T\Delta S$ [equation (4.6)].

Returning to the reaction between aluminium ions and edta^{4-} ions, we deduce that the spontaneous occurrence of this reaction must be due to a value of ΔS for the reaction being sufficiently positive so that $T\Delta S > \Delta H$. And indeed this proves to be the case, since if we go to the appropriate source and look up the value of ΔS for this reaction we find that $T\Delta S$ (at 20 °C) is about 143 kJ mol^{-1} (again this refers to standard states and is thus more properly designated as $\Delta S°$). Thus, knowing that $\Delta H°$ is 50 kJ mol^{-1}, we can readily calculate that $\Delta G°$ is $-$ 93 kJ mol^{-1}.

The relationship between $\Delta G°$ and the equilibrium constant, K_c, of the reaction is:

$$\Delta G^\circ = -RT \ln K_c \qquad (4.11)$$

where R is the gas constant ($8.314 \text{ J K}^{-1} \text{ mol}^{-1}$), and so we can calculate that K_c is about 10^{17}.

with standard edta^{4-} solution is in for a disappointment, as no sharp end-point will be observed. The reason for this is that the reaction between Al^{3+} ions and edta^{4-} ions in aqueous solution is slow compared with the normal rate of addition of titrant in a volumetric procedure. This is because the reaction must actually proceed by breaking thëaluminium–oxygen bonds in the hydrated Al^{3+} ion, and for this to happen a collision of sufficient energy with an edta^{4-} ion has to occur. The random thermal motion of the ions and water molecules in the reaction vessel produces a spread of molecular speeds, and the probability of a close encounter between an aluminium ion and an edta ion being of sufficient energy to break an Al–O bond or two is quite low.

The formal way of expressing this effect is to assign an activation energy to the reaction. This is an energy barrier that the reactants have to surmount before they will form products. For some reactions and conditions the barrier is virtually insurmountable. For example, a mixture of hydrogen and oxygen at room temperature is stable almost indefinitely despite the fact that the mixture is highly unstable thermodynamically. This reaction can be persuaded to go quite simply by locally increasing the energy of the collisions between H_2 and O_2 in order to break the H–H and O–O bonds. A lighted match will do. The energy liberated as each H_2O forms is sufficient to raise a few more H_2 and O_2 over the energy barrier, and the reaction proceeds extremely rapidly from then on.

There is no analogous procedure for the aluminium–edta reaction, although the reaction can be speeded up by heating. Nor is there a known catalyst (a device which provides an alternative reaction pathway with a lower energy barrier, usually by chemical means). The analytical chemist deals with this situation by using the technique of back-titration (see page 97).

What the analytical chemist needs to know about chemical reaction energies is summarised in Figure 4.3. The rate of a reaction depends on its mechanism. The study of the rates of reaction is known as the study of reaction 'kinetics' and is widely

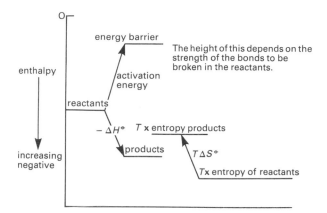

Figure 4.3　*Energies involved in a chemical reaction. The situation shown is one in which ΔH° is negative and $T\Delta S^\circ$ is positive, so, regardless of the relative magnitudes of ΔH° and ΔS°, ΔG° ($= \Delta H^\circ - T\Delta S^\circ$) is negative.*

used in chemistry as a means of investigating reaction mechanisms. Analysis is involved as well, of course, because the nature of the products has to be established, and a kinetic study will involve measuring changes in concentration (either disappearance of a reactant or appearance of a product) as a function of time. The various spectroscopic techniques described in Chapter 2 are particularly useful for this.

Before we can go on to look at titrimetric and gravimetric analysis, we need a bit more detail in the model for species in solution. Two further aspects need to be considered: the role of other ions and reactions with solvent (*i.e.* water) molecules.

Other Ions in Solution

Obviously, there are other ions in solution and, in general, it is not possible to ignore their effect. Even if they do not undergo any chemical reaction which interferes, they will form bonds with the water molecules, orientate them, introduce localised structure, and influence the behaviour of other ions by means of electrostatic interactions. The extent of these effects increases with increasing concentration. The net effect of these other ions is to make the ion we are interested in behave as though its concentration were actually less than it really is. We take this effect into account by assigning to the ion an 'activity' value which has the units of concentration but which is less than the true concentration. The ratio of activity to concentration is known as the 'activity coefficient'. An ion which is 'fully active' has an activity coefficient of 1; less active ions have values of less than 1. Most of the equations relating quantities involving properties of solutes are rigorously true only for values of the activities of species in solution. This includes equation (4.3).

An assessment of the likely effect of the other ions in solution usually starts by calculating the ionic strength, I, of the solution. This is a sort of average ionic concentration given by equation (4.12):

$$I = \frac{1}{2} \sum_{i=1}^{n} c_i z_i^2 \tag{4.12}$$

where c is the concentration, z is the charge, and n is the total number of ionic species in solution (the capital Greek sigma means 'the sum of all things like'). It is a measure of the intensity of the electric field due to the ions in solution. Providing this value is below about 10^{-4} mol dm^{-3}, activity coefficients can be assumed to be unity. Unfortunately there is no way of measuring the activity coefficient of a single ionic species (ions always come in pairs at least!), and the best that can be done is to calculate values based on a model of behaviour due to Debye and Hückel, from an equation known (not too surprisingly) as the Debye–Hückel equation. Accurate calculations of equilibrium concentrations at high ionic strengths can be quite difficult.

Reactions Involving the Solvent

Another facet of the complete picture of chemical species in solution is that some of them react with the solvent and break the bonds within the solvent molecules. In the case of water this reaction is known as hydrolysis (in general the process is called solvolysis) and the products will include, among other things, H_3O^+ and/or OH^-.

Water molecules will react with each other to a small extent, so that even if everything else is rigorously excluded from water there will still be some oxonium (H_3O^+) and hydroxide (OH^-)

ions present. In its simplest form the reaction may be expressed as $2H_2O \rightleftharpoons H_3O^+ + OH^-$, for which the equilibrium constant is given by $[H_3O^+][OH^-]/[H_2O]^2$, where the square brackets denote equilibrium concentrations. The value of the equilibrium constant is approximately 3.23×10^{-18}. As the initial concentration of water is about 55.6 mol dm^{-3}, it is not too difficult to calculate that the equilibrium concentrations of H_3O^+ and OH^- are 10^{-7} mol dm^{-3}. In turn, this means that the equilibrium concentration of H_2O is not appreciably different from the initial concentration and therefore that the concentration of H_2O can be considered a constant and incorporated into the value for the equilibrium constant (at 25 °C), giving:

$$[OH^-][H_3O^+] = K_w = 10^{-14} \text{ mol}^2 \text{ dm}^{-6} \qquad (4.13)$$

Because this term crops up quite a lot when considering what's happening in aqueous solutions, it is given the special symbol K_w (and is referred to as the ionic product of water or the self-ionisation constant). Note that K_w has the units mol^2 dm^{-6}.

It is common practice when considering equilibria involving dilute aqueous solutions to treat the concentration of H_2O as constant throughout and to incorporate this into the value of the equilibrium constant.

Some chemical species are hydrolysed (*i.e.* react with H_2O) when put in water, the result of which is to change the relative amounts of H_3O^+ and OH^- in the solution. When the concentrations of H_3O^+ and OH^- are equal the solution is said to be neutral, when H_3O^+ is in excess over OH^- the solution is acidic, and when OH^- is in excess over H_3O^+ the solution is alkaline. The extent to which one of these situations occurs depends on the initial concentrations of the species involved and the values of the equilibrium constants for the particular reactions involved. In order to calculate this we must have access to some data on values of K_c. Fortunately, extensive tables of these are available and usually the calculations will start with looking up the values of K_c for the various reactions involved. Obviously, as the number of species in the solution is increased, the number of possible equilibria increases and the calculations can become quite complicated. Sometimes simplifying assumptions can be made (such as the concentration of a particular species being so small that it can be ignored), and nowadays there are a number of computer programs that take the tedium out of the calculations.

SUMMARY

The first part of this chapter did not contain much analytical chemistry. However, the reason for the rather long introduction is that, as was stated at the beginning of this chapter, analytical chemists need to have a good grasp of some basic ideas of what is going on when chemical reactions occur (particularly those in aqueous solution). These basic ideas involving mechanism, driving forces, equilibria, and kinetics are summarised in Figure 4.4.

ANALYSES BASED ON VOLUME

Suppose you are faced with the problem of determining the concentration of nitric acid in a solution used in part of your company's manufacturing process. A convenient way of doing this is to take a portion of accurately known volume, V_H, and add accurately known volumes of an alkali solution of accurately known concentration, c_{OH}, until all the 'hydrogen' ions have reacted with the hydroxide ions. If the volume of alkali required is V_{OH}, then it is possible to calculate c_H, the unknown concentration of the acid sample solution. The procedure is summarised in Figure 4.5.

This procedure is called a titration and special equipment called calibrated glassware is used for the accurate measurement of volume. A pipette is used to transfer V_H to the titration vessel (a conical flask) and a burette is used to add alkali and to measure V_{OH}. The important design features and how to use these devices will be discussed in Chapter 5.

A number of questions concerning this procedure still remain to be answered. These are:

(1) When excess OH^- (titrant) is indicated, how much H_3O^+ (titrand) is left unreacted?
(2) How do we know when an excess of OH^- has been added?
(3) How do we know the concentration of OH^-?

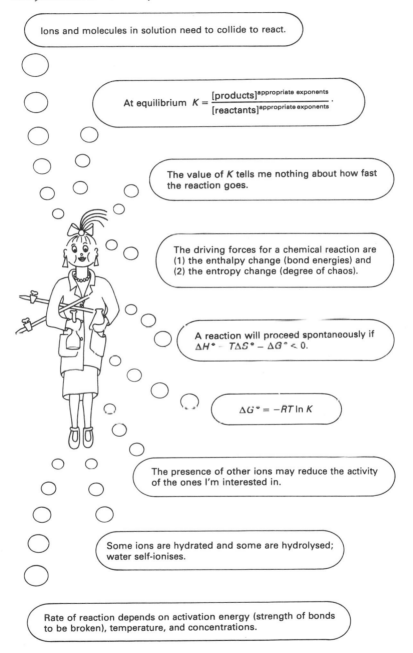

Ions and molecules in solution need to collide to react.

At equilibrium $K = \dfrac{[\text{products}]^{\text{appropriate exponents}}}{[\text{reactants}]^{\text{appropriate exponents}}}$.

The value of K tells me nothing about how fast the reaction goes.

The driving forces for a chemical reaction are (1) the enthalpy change (bond energies) and (2) the entropy change (degree of chaos).

A reaction will proceed spontaneously if $\Delta H^\circ - T\Delta S^\circ - \Delta G^\circ < 0$.

$\Delta G^\circ = -RT \ln K$

The presence of other ions may reduce the activity of the ones I'm interested in.

Some ions are hydrated and some are hydrolysed; water self-ionises.

Rate of reaction depends on activation energy (strength of bonds to be broken), temperature, and concentrations.

Figure 4.4 *All the analytical chemist needs to know about chemical reactions.*

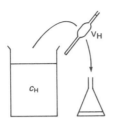

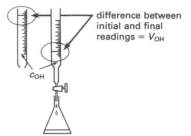

difference between
initial and final
readings = V_{OH}

c_{OH}

V_H

c_H

concentration in mol dm^{-3}
volume in cm^3

Equation for reaction tells us
the number of moles of acid
reacting per mole of alkali.

$$H^+ + OH^- \rightleftharpoons H_2O$$

Note: equation doesn't have
to be an accurate
representation of the reaction
mechanism. It just has to
show the relative proportions
of reactants reacting.

Number of moles of OH$^-$
used in titration

$$= V_{OH} \times \frac{c_{OH}}{1000}$$

i.e. volume (in cm^3) ×
number of moles per cm^3.
Equation tells us that this is
equal to the number of moles
of H$^+$ used in titration.

Therefore concentration of H$^+$

$$= \left(\frac{V_{OH} \times c_{OH}}{1000} \div V_H \right) \times 1000$$

i.e. number of moles per cm^3
× 1000, giving mol dm^{-3}.

Figure 4.5 *An acid–base titration.*

The Position of Equilibrium

In a reaction that is used for analysis we want all of the sample to react. We have to settle for some realistic value for 'all' (such as 99.9%) because the fact that equilibrium constants have finite values means that the concentration of a reactant can never be zero at the end of a reaction [see equation (4.3)]. The restrictions that this imposes are discussed in Box 11.

So far it has been assumed that the concentration of H_3O^+ in the solution for analysis is the same as the concentration of the nitric acid, *i.e.* that HNO_3 is 'completely dissociated' in aqueous solution to H_3O^+ and NO_3^-. The equilibrium constant for the reaction $H_2O + HNO_3 \rightleftharpoons H_3O^+ + NO_3^-$ is 25.1 mol dm^{-3} (in-

BOX 11   Equilibrium Considerations in Titrimetric Analysis

Suppose in the titrimetric analysis being considered here we titrate a certain volume of 0.01M H_3O^+ solution with 0.01M OH^- solution and we have a means of detecting when we have added the same volume of alkali. After reaction we want the number of moles of H_3O^+ to be no more than 0.1% of the starting number (99.9% reacted). The starting number of moles was $0.01 \times V$ (*i.e.* the concentration, in mol dm^{-3}, multiplied by the volume taken for the titration, V, in dm^3), so at the end the number of moles must be no more than $0.000\,01 \times V$ (0.1% of $0.01 \times V$). If the volume has doubled to $2V$ (equal volume of acid added), the concentration (number of moles divided by volume) must be no more than $0.000\,01V/2V$, *i.e.* 5×10^{-6} mol dm^{-3}.

We now have to find out what the concentration actually is. The problem can be reformulated as 'what happens in a solution containing 0.005 mol dm^{-3} H_3O^+ and 0.005 mol dm^{-3} OH^- when the reaction has reached equilibrium?', *i.e.* we can replace the actual experiment with a 'thought' experiment in which we first mix equal volumes of 0.01M H_3O^+ and OH^- and then allow them to react. At equilibrium $[OH^-][H_3O^+] = 1 \times 10^{-14}$ mol^2 dm^{-6} and, as equal volumes of equal concentrations were mixed, $[OH^-] = [H_3O^+]$, so $[H_3O^+]$ is 1×10^{-7} mol dm^{-3}. As this is 50 times lower than the value required for 99.9% reaction, the reaction will form the basis of a useful analytical method. The limitations on reactions for titrimetric analysis now begin to become apparent. Firstly, the value of the equilibrium constant, K_c, has to be sufficiently large that, for the initial concentrations involved, 99.9% reaction is achievable. Secondly, for a given value of K_c there is a limit to how low the initial concentration can be for 99.9% reaction to be achieved still.

corporates [H_2O]). This means that if 0.01M HNO_3 'dissociates' then at equilibrium the concentration of H_3O^+ is 0.009996 mol dm^{-3}, *i.e.* the nitric acid is 99.96% dissociated, and we can therefore take the concentration of H_3O^+ to be equal to the concentration of HNO_3.

Solutions of acids which are almost completely dissociated are known as 'strong' acids. The term is also applied to other solutions containing 'completely' ionised material. Thus we have strong bases (*e.g.* sodium hydroxide solutions) and strong electrolytes (*e.g.* potassium chloride). Electrolyte is a general term for a conducting solution. The effect of the strength of the acid on the titration is explained in Box 12.

BOX 12   Effect of Acid Strength

Let's see what happens when we replace the strong acid in our analysis with a weak one, ethanoic acid. The equilibrium constant for the dissociation $H_2O + CH_3COOH \rightleftharpoons CH_3COO^- + H_3O^+$ is 3.13×10^{-7}. For a reason which was explained earlier in connection with K_w (see page 81), the value of $[H_2O]$ is considered constant and incorporated into the numerical value of the equilibrium constant. These acid dissociation constants are usually given the symbol K_a (the corresponding equilibrium constant for a base is referred to as K_b) and have units mol dm^{-3}. In this particular case K_a has the value 1.74×10^{-5} mol dm^{-3}. Values of equilibrium constants are often very small, depending on which way round the equation is written. The number of zeros either before or after the decimal point can be rather cumbersome, so a special way of expressing the number is used rather than just the normal standard form (one digit to the left of the decimal point multiplied by 10 to the appropriate power). This is the 'p' notation, which stands for the 'negative logarithm (to base 10) of'. This is equivalent, of course, to 'logarithm (to base 10) of reciprocal of'. Thus for ethanoic acid the pK_a is 4.76. The notation is sometimes used for expressing the concentrations of chemical entities in solution, particularly the oxonium ion concentration, which is very often expressed as the appropriate pH value, the 'H' standing for 'H$^+$' since the term was coined before it was considered that 'H$^+$' existed mainly as H_3O^+ in aqueous solution. As oxonium ion concentrations can vary across very wide ranges, the pH scale has a convenient set of numbers to work with.

In the titration of ethanoic acid with sodium hydroxide the overall reaction being used for the analytical method is $CH_3COOH + NaOH \rightleftharpoons CH_3COONa + H_2O$. However, when calculating the concentrations of CH_3COOH *etc.* at the point when an equivalent amount of sodium hydroxide has been added, we have to bear in mind that the other equilibrium, $H_3O^+ + OH^- \rightleftharpoons 2H_2O$, has to be satisfied. The concentrations in solution of undissociated NaOH and CH_3COONa are considered to be so small that they can be neglected. When an equivalent amount of sodium hydroxide has been added, the concentration of CH_3COOH has decreased to 1.69×10^{-6} mol dm^{-3}, *i.e.* about 99.97% reaction. However, the oxonium ion concentration is 5.89×10^{-9} mol dm^{-3} (pH = 8.23) whereas in the previous titration the pH was 7.00 at the end of the titration.

As the acid being titrated gets weaker (*i.e.* the pK_a gets larger), this trend continues, *i.e.* the pH at the equivalence point increases and also the percentage reacted becomes smaller. This is summarised in Figure 4.6. The calculations are not particularly difficult but are tedious and involve the solving of cubic equations. A number of simplifying

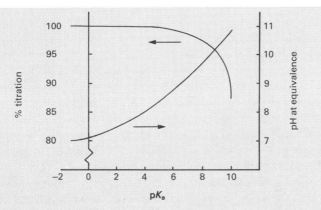

Figure 4.6 *Relationship between pK_a, % titrated, and pH at the equivalence point.*

assumptions can be made (*e.g.* neglect concentrations which are very much smaller than others in some of the equations). The whole business is much easier if you have the appropriate software for your computer.

The trends shown in Figure 4.6 are fairly readily explained in terms of what we know is going on in the solution. As the pK_a of the acid increases, less and less acid is dissociated. The effect on the pH at the equivalence point can be visualised by considering the titration of the weak acid in two stages (another 'thought' experiment). The first stage is the same as for a strong acid, so that the concentration of H_3O^+ at the equivalence point is 10^{-7} mol dm^{-3}; in the second stage, however, the anion of the acid (known as the conjugate base) reacts with H_3O^+ to form the undissociated acid. This upsets the $H_3O^+ + OH^- \rightleftharpoons 2H_2O$ equilibrium, so some water self-ionises to maintain the equilibrium concentration product, $[H_3O^+][OH^-]$, at 1×10^{-14} mol^2 dm^{-6}, increasing $[OH^-]$ over $[H_3O^+]$ and making the solution alkaline. The more H_3O^+ is removed in this second stage (*i.e.* the bigger the pK_a value), the lower is $[H_3O^+]$ and the higher the pH. The percentage of acid reacted obviously decreases as more undissociated acid is left at the equivalence point.

Solutions containing incompletely ionised species are known as 'weak' electrolytes. There does not seem to be any hard and fast rule for deciding what 'not completely' means in terms of a value of K_c, but about 0.0001 mol dm^{-3} appears to be a guide-line. It should be noted that the terms 'strong' and 'weak' do not refer to the concentration of the solution. They are being used in a strict chemical sense and not in the everyday sense where 'strong' means 'concentrated' and 'weak' means 'dilute'. It is perfectly possible for a strong electrolyte to be a very dilute solution.

In a real analysis, of course, the initial concentration of the acid is not known and some way of recognising when the equivalence point has been reached is needed. To do this, we have to recognise when we have an excess of titrant, and so the concentration of OH^- in the titration vessel is monitored. Let's assume that we have some method of doing this. In the case of a strong acid the titrant was in excess above pH 7, so let's take as an indication that the titration is over to be when the pH just increases above 7.00. In the case of ethanoic acid the pH rises above 7.00 after 98.75% of the volume required for equivalence has been added. If $pK_a = 8$, the end-point is indicated when about 10% of the volume for equivalence has been added, and the situation is even worse for $pK_a = 10$!

Obviously, analytical chemists must know something about the pK_a of the acid being titrated so as to set the appropriate value of the pH to take as the end-point. The more accurately this is known and the more accurately the concentration of OH^- during the titration can be followed, the more closely the end-point (as indicated when the pH gets above a certain value) will coincide with the equivalence point. This is what is wanted if the titration is to give accurate results.

How do we in fact monitor the concentration of OH^- during a titration?

End-Point Detection

In Chapter 3 a potentiometric device which responded to the concentration of H_3O^+ was described, the so-called pH electrode. Although the electrode does not respond strictly to the hydrogen ion concentration, and the accurate conversion of the millivolt output of the meter to H_3O^+ concentration values is quite a problem, the device is useful for following the course of a titration. In the case of the titrations of acid with alkali, we want to follow the OH^- concentration, but, as the concentrations of H_3O^+ and OH^- will always satisfy the condition $[H_3O^+][OH^-] = 10^{14} \text{ mol}^2 \text{ dm}^{-6}$, the concentration of OH^- can easily be calculated from a pH value. As the pH of the solution is such a widely used concept, it is normal to use pH values even when the concentration of OH^- is greater than that of H_3O^+.

Suppose we have 20 cm^3 of 0.01M acid in the titration vessel and we titrate this with 0.01M sodium hydroxide, how does the pH change during the titration and what happens after the equivalence point? Again the calculations are not difficult, but are tedious, and it is much easier if you have the appropriate software for your computer. The results of the calculations for various pK_a values are shown in Figure 4.7. Now it can be seen that not only

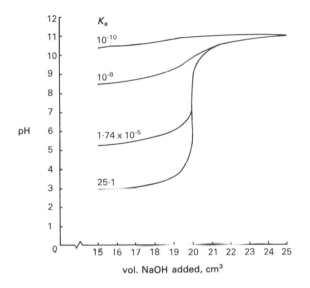

Figure 4.7 *Variation of pH during the titration of 20 cm^3 of acids of different K_a with sodium hydroxide. Note the decrease in the extent of the pH 'break' as K_a decreases.*

does the pH at the equivalence point increase as pK_a increases but also the change in the pH around the equivalence point decreases dramatically. Thus we need to know not only the pH at the equivalence point but also how sharp the 'break' in the curve of pH versus volume of titrant is going to be. For acids as weak as p$K_a = 10$ no titration is possible since, although we know that the pH at the end-point is 10.82, the pH only changes from 10.73 to 10.93 when between 19 and 21.46 cm^3 of NaOH is added. Given that with a real pH electrode there will be a certain amount of uncertainty in the value measured, it will be impossible to identify the volume at which the pH is 10.82 exactly. Contrast this with the case for $pK_a = -1.40$, where the pH changes from 5.6 to 8.4 as the volume of NaOH added goes from 19.99 to 20.01 cm^3.

As the acid being titrated becomes progressively weaker, the pH 'break' becomes harder to detect. There are a number of mathematical 'dodges' that can be used to help identify the equivalence point. The curve (Figure 4.7) obtained can be differentiated and the point of maximum slope identified, or the original 'pH versus volume of NaOH' function can be transformed to give linear functions for the regions before and after the equivalence point, whose intersection locates the equivalence point.

Buffers

Although the very weak acid solution could not be analysed by titration with sodium hydroxide solution, solutions of such materials are very useful in certain analytical techniques because they are *resistant* to changes in pH. Although quite a lot of strongly alkaline solution is being added, the pH only changes by about 1 unit between 15 cm^3 and 30 cm^3 of NaOH added. Such solutions are called buffer solutions and are very useful in controlling the 'hydrogen' ion concentration in reactions where a change in this concentration would interfere with the main analytical reaction (a good example is metal complexation reactions; see page 96).

The ethanoic acid titration has a buffer region well before the equivalence point. The solution has greatest capacity to resist pH changes when it is half neutralised. The composition of the solution at this point is the same as would be obtained by mixing equal amounts of sodium ethanoate and ethanoic acid. The pH is given by the simple relation $pH = pK_a$. Thus an ethanoate buffer works best at pH 4.76.

Visual Indicators

For most acid–base titrations a pH electrode is not used to follow the titration, but a small amount of an indicator is added. This is a strongly light-absorbing species whose structure and hence absorption spectrum depend on the pH. Generally the indicators used are weak acids or very weak acids and so the indicator is being titrated as well as the component of interest. However, the concentration of indicator in the solution is about 10^{-5} mol dm^{-3} and so the volume of sodium hydroxide equivalent to the indicator is very small, 0.02 cm^3 in the example being used here. The different colours of the indicator correspond to whether one form or another predominates in solution. Taking methyl orange as an example, the equilibrium which gives rise to the colour change is shown in Figure 4.8.

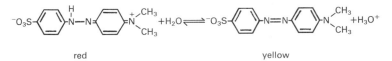

red   yellow

Figure 4.8 *Methyl orange. The colour of the solution depends on the pH.*

In general the indicator reaction can be abbreviated to HIn + $H_2O \rightleftharpoons In^- + H_3O^+$; the p$K_a$ for this reaction is 3.7, *i.e.* $K_a = 2.0 \times 10^{-4}$ mol dm^{-3}. A colour change will be seen when the concentration ratio of the red form to the yellow form changes from 10 : 1 to 1 : 10. This can be converted to a change in pH of from pK_a − 1 to pK_a + 1, by considering the equilibrium of the indicator forms. Thus, if an acid solution containing methyl orange were titrated with sodium hydroxide, a colour change from red to yellow would be observed over the pH range 2.7 to 4.7. In fact the range observed may be a bit narrower.

As has been pointed out already, the amount of OH$^-$ involved in the equilibrium with the indicator is very small. The volume of titrant required to cause a colour change is even smaller and so, compared with the volume required for reaction with the main analyte species, it can be neglected. Thus it is possible to assess the usefulness of an indicator by calculating (or observing) the pH range over which a colour change occurs and comparing this with the pH 'break' around the equivalence point in the titration. The pH ranges for two indicators, methyl red and phenolphthalein, are shown in Figure 4.9 on the titration curves for nitric and ethanoic acids with a strong-base titrant solution.

If methyl red were used as the indicator, it would be suitable for the titration of HNO$_3$ since when the indicator had turned completely yellow (pH 6.3) the volume added would be within 0.01 cm^3 of the equivalence volume. But for ethanoic acid the volume added would be 0.6 cm^3 short of the equivalence volume, and the end-point would significantly underestimate the equiva

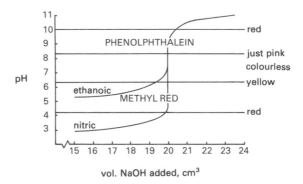

Figure 4.9 *Range of pH values for colour changes of two indicators and titration curves for 20 cm³ of ethanoic and nitric acids with sodium hydroxide.*

lence point. Below pH 8.3 phenolphthalein is colourless, so the first detectable pink colour depends on the concentration of the indicator. If this indicator were used, and the first pink tinge taken as the end-point, the nitric acid titration would have been overshot by about 0.01 cm³, but if a deep-pink colour was used as the end-point then this would overestimate the equivalence point by about 0.45 cm³. The corresponding errors for the titration of ethanoic acid would be about the same.

So, if analytical chemists want to get the right answer from an acid–base titration, they must know about the equilibria in the solution and the effect that the pK_a has on the change of the concentration of H_3O^+ (or OH^-) during the titration and on the factors to be taken into account when choosing an indicator and deciding which colour to use as the end-point (just started to change or completely changed?).

How Do We Know What the Concentration of Titrant Is?

With some substances used in titrimetric analyses it is possible to weigh out a certain amount of the chemical, dissolve it, make it up to volume in a graduated flask, and thereby produce a solution whose concentration is known accurately. This means that the chemical concerned must be very pure and stable under the storage conditions used. If the purity is $100 \pm 0.02\%$ then the chemical is known as a 'primary standard'.

In acid–base titrations sodium carbonate, potassium hydrogen phthalate, and disodium tetraborate–10-water (borax) are used as primary standards. Sodium hydroxide is not obtainable at the required purity and so after dissolution of the appropriate amount the resulting solution must be 'standardised' before use by titrating against a standard solution (of, say, potassium hydrogen phthalate). Precautions must be taken to avoid errors due to the dissolution of CO_2 in the solution, as hydroxide ion solutions will take up a lot of CO_2 due to the reactions $CO_2 + 2H_2O \rightleftharpoons HCO_3^- + H_3O^+$, $\quad HCO_3^- + OH^- \rightleftharpoons CO_3^{2-} + H_2O$, $CO_2 + H_2O \rightleftharpoons H_2CO_3$, $\quad$ and $\quad H_2CO_3 + 2OH^- \rightleftharpoons CO_3^{2-} + 2H_2O$.

The titration of bases is briefly discussed in Box 13.

BOX 13   Titration of Bases

The titrations of various acids with a strong base have been used as examples. Exactly similar arguments can be applied to the titrations of bases with acids, only this time the pH is decreasing steadily and the appropriate decisions must be taken about how the colour changes of the indicators chosen relate to the end-point. However, not all bases contain ionisable OH groups as does sodium hydroxide. Ammonia solution, for example, is basic because the ammonia hydrolyses according to $NH_3 + H_2O \rightleftharpoons NH_4^+ + OH^-$. The equilibrium constant for this reaction is sometimes referred to as the base dissociation constant, and given the symbol K_b, and sometimes as the hydrolysis constant, K_h. However, acids and bases can be put on a common footing by considering the equilibrium $NH_4^+ + H_2O \rightleftharpoons NH_3 + H_3O^+$ and quoting the equilibrium constant for this as an 'acid dissociation' constant. For this reaction the value of pK_a is 9.25, from which it is easy to calculate that the pK_b for ammonia is 4.75 because $K_a \times K_b = K_w$, *i.e.* $pK_a + pK_b = 14$. It is the relative values of pK_a and pK_b for a chemical in aqueous solution that govern whether we label it an acid or a base. If pK_a is less than pK_b, the substance is an acid; if pK_b is less than pK_a, the substance is a base.

Other Types of Titrations

We have spent some time considering the basic properties of acid–base titrations and the underlying chemical principles that analytical chemists need to understand if they are to use these reactions to obtain useful analytical information. This is not the only type of reaction that can form the basis of a titrimetric method, but the principles established for acid–base reactions also hold for other types of reaction. In other types of titration the water doesn't play quite such an interactive role as it does in acid–base reactions. There are two main groups of reactions used: oxidation–reduction (or redox) reactions and metal–ligand complexation reactions.

Redox Titrations

Most of the reactions of this type that are useful analytically have very large equilibrium constants. For example, in the oxidation of iron(II) to iron(III) in acid solution by tetraoxomanganate(VII) [which is reduced to manganese(II)]:

$$MnO_4^- + 5Fe^{2+} + 8H_3O^+ \rightleftharpoons Mn^{2+} + 5Fe^{3+} + 12H_2O \quad (4.14)$$

the value of K_c is 10^{63} dm^{-12} mol^{-4} (pK_c = −63), showing the equilibrium to be well over to the right of equation (4.14). As MnO_4^- is intensely coloured, there is no need to add a separate indicator, and the first excess of MnO_4^- gives a pink tinge to the solution. Some redox titrations do require indicators that are oxidised or reduced by the titrant in order that the end-point can be located. These function in a similar manner to the pH indicators used in acid–base titrations, and similar criteria apply in choosing an indicator for a particular titration and deciding what extent of colour change should be used to indicate the end-point. For example, 'ferroin', the complex between iron(II) and 1,10-phenanthroline, is used to indicate the end-point in the titration of arsenic(III) with cerium(IV). The reduced form, ferroin, contains iron(II) and is red, whereas the oxidised form, ferriin, contains iron(III) and is very pale blue.

This particular titration requires a catalyst to make it go fast enough to be useful; osmium tetroxide is used. The titration of arsenic(III) is often used to standardise oxidising agents, as arsenic(III) oxide can be obtained in a very pure state. Careful handling of both the solution and the solid is necessary since arsenic is toxic. Standardisation is often necessary with redox titrations as it is difficult to store solutions of strong oxidising or reducing agents. They have a tendency to react with the water or any other species in the solution, and reducing agents tend to react with atmospheric or dissolved oxygen.

Applying a redox reaction analytically may involve the analytical chemist in some pretreatment of the sample to get the analyte species into the right oxidation state (see Box 14).

BOX 14   Sample Pretreatment in Redox Titrations

In the determination of the iron content of an iron-ore sample the dissolution procedure will almost certainly involve oxidation (H_3O^+ is an oxidising agent, as are several anions of commonly used acids) and the iron will be in solution as Fe^{3+}. The solution is then passed down a column of solid reducing agent (such as zinc amalgam), which reduces the Fe^{3+} to Fe^{2+}. The solution can now be titrated with tetraoxomanganate(VII). Copper can be determined in brass by first dissolving the sample to give Cu^{2+}, in solution, then adding iodide, I^-, to reduce the copper to Cu^+, and precipitating it as Cu_2I_2 with the formation of an equivalent amount of iodine, I_2:

$$2Cu^{2+} \text{ (aq)} + 4I^- \text{ (aq)} \rightleftharpoons Cu_2I_2 \text{ (s)} + I_2 \text{ (aq)} \qquad (4.15)$$

The iodine can be titrated with thiosulphate:

$$I_2 + 2S_2O_3^{2-} \rightleftharpoons 2I^- + S_4O_6^{2-} \qquad (4.16)$$

The indicator for reaction (4.16) is usually starch, which forms an intense blue colour with iodine. Unlike most other titrations, the indicator is added just before the end-point is reached (*i.e.* when the colour of the solution due to the iodine is 'pale straw'). The reduction of Cu^{2+} to Cu^+ by I^- is quite an interesting reaction because the equilibrium for the hypothetical reaction (4.17):

$$2Cu^{2+} + 2I^- \rightleftharpoons 2Cu^+ + I_2 \qquad (4.17)$$

lies well to the left (as written). It is only because the product Cu^+ is continuously removed from solution by precipitation that the reaction is pulled over to the right (as written).

Because redox reactions involve electron transfer, it is possible to follow a redox reaction using an indicator electrode. This is just a piece of platinum whose potential is related to the relative concentrations of the oxidised and reduced species in the solution. The reaction is followed by measuring the potential difference between the indicator electrode and a reference electrode (such as the saturated calomel electrode; see page 59). The shape of the plot of potential difference against volume of titrant added is very similar to the pH curves obtained with acid–

base titrations. A large potential 'break' occurs around the equivalence point. This is a useful way of following a redox reaction to locate the end-point.

It should always be borne in mind that the equilibria involved in redox titrations may depend on the pH of the solution as H_3O^+ or OH^- may be involved in the reaction.

Complexometric Titrations

There is only one complexing ligand of any importance in this class of titrations and that is 1,2-bis[bis(carboxymethyl)amino] ethane (ethylenediaminetetra-acetate or edta). The reason why edta forms such strong complexes with metals has already been discussed earlier in this chapter. Edta forms complexes with a large number of metals (virtually all except the alkali metals); regardless of the size and charge of the metal ion, the complexes are always between one metal ion and one edta ligand.

Conditions?   The pH of the solution is very important. As with other reagents forming metal complexes, the ligand will react with H_3O^+ (H^+ is attached to the co-ordination sites), and if the concentration of H_3O^+ is too large (decreasing pH) the 'protons' may win the competition for the ligand over the metal. It all depends on the relative values of the equilibrium constants for the protonation reaction and the complex-forming reaction. It may therefore be necessary to buffer the solution so as to keep the pH constant during the titration.

Suppose the pH of the edta titrant is 4. At this pH edta exists almost entirely as H_2edta^{2-}, and so in a complexation reaction with a metal 2 moles of H_3O^+ will be formed per mole of metal ion titrated. Thus if a 0.01M solution of a metal is titrated with a 0.01M edta solution the increase in the concentration of H_3O^+ at the equivalence point would be 0.01 mol dm^{-3} if the reaction went to completion. However, the decrease in the pH may prevent the reaction from going to completion and the end-point underestimates the equivalence point. Above pH 11 edta is almost entirely present as edta^{4-} and some metals can only be titrated at alkaline pH values. A lot of metals will precipitate as hydroxides in alkaline solution and so they have to be kept in solution by another complexation reaction, such as the reaction with ammonia to form ammines. Ammonia is quite often a constituent of a buffer solution at pH 10, so the use of an ammonia buffer keeps the pH at an appropriate value and the metal in solution as a soluble ammine complex.

Indicators. The visual indicators used are known as metallochromic indicators because they have different colours depending on whether they are complexed with a metal or not. The indicator forms a weaker complex with the metal than does edta, so at the beginning and during the titration the colour seen corresponds to the metal–indicator complex. The edta first titrates all the free (hydrated) metal ions, but as the end-point is approached the metal complexed with the indicator reacts with the edta, releasing the free indicator with a consequent change in colour.

As with acid–base and redox indicators, a little thought is required in choosing the indicator. The metal–indicator complex must not be so stable that a considerable excess of edta must be added for a colour change to be seen, nor too weak that a gradual change in colour is seen prior to the equivalence point. Some thought must also be given to the acid–base behaviour of the indicator as it won't be much good if, at the pH of the reaction, the free indicator is the same colour as the metal–indicator complex.

Modus Operandi. There are several ways in which edta titrations can be applied, particularly when dealing with mixtures of metals.

Standardisation. Solutions of edta are normally made from the disodium salt as it is sufficiently soluble in water (unlike the acid itself) and doesn't produce the very alkaline solutions that the tetrasodium salt does. High-purity zinc dissolved in nitric acid may be used to standardise the edta solution.

Back-Titrations. Some metal ions, such as chromium(III) and aluminium, react rather slowly with edta. There are a variety of reasons for this, including, in the case of Cr(III), the considerable stability of the electronic structure of the ion in an octahedral co-ordination environment and, in the case of aluminium, the strong metal–oxygen bonds of the stable hydrated ion. Aluminium, as do some other metal ions, exists in aqueous solution as a polymeric species with oxygen bridges between the metal atoms. This means that only the terminal aluminium is available for reaction with edta. When the reaction between the metal ions and edta is slow, the approach is to use what is known as a 'back-titration'. A known amount of edta, in excess of that required to react with the metal ion, is added and the solution is heated long enough for the edta–metal ion complex to form. The amount of unreacted edta is determined by titration with a metal ion that forms a weaker complex with edta than the metal sought. If the back-titration complex were more stable, the back-titrant metal ion would displace the first metal ion from the edta complex already formed.

pH Control.   It is possible to analyse for more than one metal in the same solution if the K_c values for the metals are sufficiently different to allow the pH to be adjusted so that only one metal is titrated. After titration of the first (in the most acidic conditions), the pH is raised to that at which the second may be titrated. In theory, three metals could be titrated sequentially in this way. In practice, by the time the titrant and buffers have been added the volume of solution gets rather large and unwieldy after two metals have been titrated and the indicator colour change becomes rather faint and diffuse.

How do we know the pH values required? Actually, this is no problem for the analytical chemist who understands what's going on in the solution, as no other indicators are needed for the adjustment of the pH for the following reasons:

(1) If, on adding the metallochromic indicator, the free indicator colour forms, the pH is too low to titrate any of the metals (even if edta will react, the indicator is not working), so go to paragraph (3).

(2) If, on adding the metallochromic indicator, the metal indicator colour forms, the pH is high enough for edta to titrate the metal (the indicator complexes are weaker than the edta ones, so if the indicator reacts with metal then edta will as well). But the pH might be so high that a second metal is titrated or partially titrated as well, so acid is added until the free indicator colour appears.

(3) Then the pH is slowly raised (by dilution possibly) until the metal indicator colour just appears. Now the pH is correct for titrating the first metal.

(4) After titrating the first metal the pH is raised until the free indicator colour (end-point of first titration) changes to the (second) metal indicator colour, and after buffering at this pH the

second metal may be titrated. It works very nicely for the titration of bismuth at pH 1 followed by lead at pH 5 (hexamine buffer), using xylenol orange as indicator.

Masking. There are other ways of dealing with mixtures. These are based on the reactions that some metals undergo with other complex-forming reagents. If a ligand is added which forms a stronger complex with the metal than does edta, this metal will not be titrated when edta is added.

For example, iodide forms a stronger complex with mercury than does edta, so a method for determining mercury and another metal can be based on this. The most convenient way of doing this is by back-titration. Consider the determination of mercury and cadmium. A known amount of standard edta is added and the excess (not reacted with Hg or Cd) is titrated with standard zinc. Iodide is added which releases edta from the Hg–edta complex but not from the Cd–edta complex. The released edta is titrated with the standard zinc solution. The first part of the determination tells us how much Hg + Cd is present and the second part how much Hg is present. The amount of cadmium can be determined by difference.

ANALYSES BASED ON MASS

In a gravimetric analysis (as an analysis based on weighing the product of a reaction is known) the species being determined may be precipitated from the sample solution as an insoluble compound of known composition. The precipitate is transferred to a pre-weighed sintered glass crucible (solid material is trapped on a disc of sintered glass, see page 113) and washed, and the crucible and contents are dried and re-weighed. The overall procedure is outlined in Figure 4.10.

Chemical Background to Solubility

The processes of precipitation and dissolution are governed by the same considerations of enthalpy and entropy as are homogeneous (all reactants and products in same phase) reactions. Contributions to the overall enthalpy change include bond energies in the solid, the solvation energy of the resultant ions in solution, and the energy needed to 'make a hole' big enough in the solvent into which to put the ion. Entropy effects will include

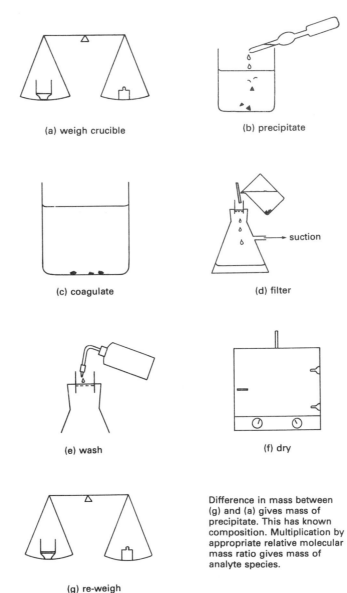

(a) weigh crucible

(b) precipitate

(c) coagulate

(d) filter

(e) wash

(f) dry

(g) re-weigh

Difference in mass between (g) and (a) gives mass of precipitate. This has known composition. Multiplication by appropriate relative molecular mass ratio gives mass of analyte species.

Figure 4.10   *Outline of gravimetric analysis.*

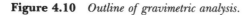

the loss of the solid crystal structure as its components are dispersed in solution and the ordering of the solvent structure due to the solvation of the ions.

The net balance of these effects can be quite finely poised and the sign of the Gibbs free-energy change difficult to predict. For example, sodium chloride and barium chloride are soluble in water whereas silver chloride and barium sulphate are insoluble. However, as a general rule, a chemical species may be soluble in water if it can ionise or can be ionised to give charged entities which can be solvated or if the 'centre of gravity' of the positive charge is spatially separated from the 'centre of gravity' of the negative charge and the molecule is polar. If the solid can't form ions or is not polar, it is unlikely to be soluble, and, if the ingredients for the formation of the solid can be brought together in aqueous solution, a precipitate will form.

This is the basis of a number of gravimetric methods for the determination of metals. For example, nickel can be precipitated by butanedione dioxime (dimethylglyoxime, dmg) as shown in Figure 4.11.

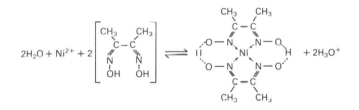

Figure 4.11 *Reaction between nickel and dimethylglyoxime. The dotted lines in the complex indicate that each hydrogen is shared by the two oxygens.*

Dmg is not very soluble in water, as you may deduce from the above discussion, but ionisation of the OH groups gives the molecule just enough interaction with the solvent to get sufficient into solution to react with the nickel.

Unlike edta, which forms useful complexes with a wide range of metals, dmg only precipitates nickel and palladium (although it does form soluble complexes with many other metals). It is the interaction between metal ions that is important. Extra stability is incurred in the crystal structure of the solid $Ni(dmg)_2$ complex because of the formation of Ni–Ni bonds. The $Ni(dmg)_2$ molecule is essentially flat, and the Ni–Ni bonds are formed above and below the plane of the molecule.

Conditions

As with complexometric titrations, complexometric precipita-
tions will only be successful if the conditions are right, and thus
the pH must be controlled. For example, if the above nickel
solution is too acid, the complex will not precipitate owing to the
preferential protonation of the oxygens of the dmg. As oxonium
ions are formed as the precipitation proceeds, the solution should
be buffered to avoid a drop in pH otherwise precipitation will be
incomplete. Care should be taken that a component of the buffer
solution doesn't form a strong complex with the metal under
study as this may mask the metal ion. The formation of soluble
complexes can, of course, be used to prevent interference by
another metal.

Equilibrium Constants

The equilibrium constant for the reaction aA (aq) + bB (aq) $\rightleftharpoons$
cC (s), for which $K = [C]^c/[A]^a[B]^b$, has to be above a certain value
in order that the amount of the analyte (species being determined) left in solution at the end of the reaction is below a certain
proportion of the starting amount, say 0.1%. As the product, C, is
a solid, its concentration is a constant (concentration is amount
per unit volume, which for a solid is proportional to the density),
and this is incorporated into the value of K. For the reaction
written as above the equilibrium constant is known as the
'insolubility product', K_{ins}. If the reaction were considered in the
opposite direction, cC (s) $\rightleftharpoons$ aA (aq) + bB (aq), the equilibrium
constant is known as the solubility product, $K_{sp} = [A]^a[B]^b$.
Obviously $K_{ins} = 1/K_{sp}$. It is common for values of the solubility
product to be given in tables of data (but not of the insolubility
product).

Gravimetric methods differ from titrimetric methods in two
important aspects. Firstly, with a gravimetric method it is possible
to add an excess of the precipitating agent to decrease the amount
of unprecipitated analyte left in solution at the end of the
experiment. There may be a limit to this excess if the precipitating agent itself is not particularly soluble (like dmg), as a false high
result may be obtained if the required precipitate contains some
precipitated reagent. Secondly, an indicator is not necessary and
thus there are no problems analogous to those of deciding when
the equivalence point has been reached because, from the first

point above, gravimetric methods are always 'overtitrated' to improve their accuracy by ensuring completeness of precipitation.

Applications

Gravimetric methods will be used in situations where (a) the highest accuracy ($\pm 0.1\%$) is required and (b) there is sufficient analyte (>10 mg) to be collected and weighed after precipitation. Gravimetric methods are thus not suitable as finishes for trace analyses. In addition they are slow and need skilled analytical chemists not only from the point of view of minimising the manipulative errors but also for designing a method to solve a new problem. In such a situation analytical chemists must use all their chemical knowledge to devise a method which will precipitate the analyte species quantitatively (more than 99.9%) but which will not suffer from interferences from other components of the mixture.

However, despite the increasing use of instrumental finishes for analytical methods, analytical chemists still need to have both of these skills because (a) as gravimetric methods are the most accurate, it may be necessary to check a more rapid method, for use in a routine situation, against a gravimetric method from time to time and (b) an overall method may well need to include a precipitation step either to remove an interference or as a method of preconcentration.

FURTHER READING

P.W. Atkins and M.J. Clugston, 'Principles of Physical Chemistry', Pitman Publishing, London, 1983.

P.A.H. Wyatt, 'The Molecular Basis of Entropy and Chemical Equilibrium', The Royal Institute of Chemistry, London, 1971.

P.A.H. Wyatt, 'A Thermodynamic Bypass GOTO log K', The Royal Society of Chemistry, London, 1982.

P.G. Ashmore, 'Principles of Reaction Kinetics', 2nd Edn., The Chemical Society, London, 1973.

D.O. Cooke, 'Inorganic Reaction Mechanisms', The Chemical Society, London, 1979.

D.T. Burns, A. Townshend, and A.H. Carter, 'Inorganic Reaction Chemistry', Ellis Horwood, Chichester, 1981, Volume 2 (Reaction of the elements and their compounds), Parts A and B.

R.A. Chalmers, 'Aspects of Analytical Chemistry', Oliver and Boyd, Edinburgh, 1968.

D. Betteridge and H.E. Hallam, 'Modern Analytical Methods', The Chemical Society, London, 1972.

O. Budevsky, 'Foundations of Chemical Analysis', Ellis Horwood, Chichester, 1979.

D.C. Harris, 'Quantitative Chemical Analysis', W.H. Freeman and Co., San Francisco, 1982.

Chapter 5
Tools of the Trade

In some respects professional chemists are like doctors or dentists: as well as having an intellectual grasp of the subject, they must possess certain manual skills. Not only must chemists be able to use the tools peculiar to their own particular trade, but they must also be able to use certain tools common to a number of scientific disciplines, such as computers, and some mathematical techniques and statistical methods.

The chemist's own particular tools will include the various pieces of glassware *etc.* used for the manufacture, purification, and analysis of materials, as well as a variety of instruments such as pH meters, balances, spectrometers, *etc.* The analytical chemist must be skilled at using all these tools (particularly those involved in quantitative analytical work) and, like any good craftsman, must know what a particular tool can do and when it is appropriate to use it. This is one of the reasons why chemistry courses contain a large amount of practical work. As well as providing an opportunity to see and handle chemicals, to 'illustrate' theory, to obtain a number for a physical constant, to analyse a material, or to synthesise a compound, experimental work is designed to teach students to be skilled with their hands and in their choice and use of tools. To become skilled requires practice, just as a musician has to practise, and so there will be an element of repetition in practical work to allow students to acquire these skills. Fortunately, chemistry is sufficiently varied for a well designed practical course to incorporate lots of different chemistries and so remove some of the tedium associated with the acquisition of skills by repetition.

No doubt you would feel unhappy about consulting a surgeon

or a dentist who had not received a proper training in practical skills, and chemists should view their own professional standing in the same way. If you are going to call yourself a chemist, you must be skilled in using the tools of chemistry as well as having an understanding of its theories.

QUANTITATIVE MANIPULATIONS

Many of the skills that the analytical chemist needs to possess centre around being able to transfer a known volume of liquid or to handle 100% of the precipitate produced by a chemical reaction.

Calibrated Glassware

Competent analytical chemists will be aware of the following methods of working with calibrated glassware and of the associated reasons for these operating procedures.

Flasks.   The basic 'tool-kit' of the analytical chemist will contain, among other things, a set of stoppered flasks specifically made to hold a certain volume. A typical example is shown in Figure 5.1(a). The neck of the flask is narrow so that small changes in the liquid volume are observed readily as changes in the level of liquid in the neck. The flask will contain the nominal volume of

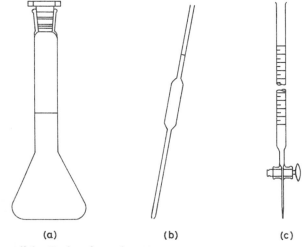

(a)   (b)   (c)

Figure 5.1   *Basic volumetric apparatus.*

liquid when the meniscus just 'touches' the circular graduation mark around the neck when the mark is viewed at eye level. Calibrated, 'graduated', or 'volumetric' flasks must be treated with care and not heated or cooled to any great extent. Glass expands and contracts very slowly. As with all items of glassware, flasks must be kept scrupulously clean. Soaking in an appropriate detergent solution (there are several commercially available preparations specifically designed for laboratory use) is carried out to remove grease. Coloured solutions should be kept in flasks for the minimum period of time, especially if the colour is due to an organic chemical which could adsorb onto and thus stain the glass. The chemical properties of glass should not be forgotten: it will dissolve in alkaline solution and act as an ion exchanger by removing cations from solution and exchanging them for sodium, potassium, *etc*. This may be important when conducting trace analytical work.

Calibrated flasks should not be used for storing solutions. Cheaper polythene (or similar) bottles are used instead and the flask is rinsed out ready for re-use. In some trace analytical work it may be necessary only to use one particular flask for a particular solution, to avoid cross-contamination. For example, it will almost certainly be disastrous for the analysis if the flask in which a 10 000 mg dm^{-3} metal solution was prepared is re-used later in the method to prepare a 10 ng dm^{-3} solution. Some analyses require the use of solutions which attack glass, such as hydrofluoric acid, and in these cases flasks made of Teflon or a similar inert fluorocarbon are used.

When filling a graduated flask to the mark, the last fraction of a cubic centimetre is added by carefully running the liquid down the neck using a dropping pipette or by careful addition from a wash bottle, while holding the flask so that the graduated mark is at eye level. Leaving the flask on the bench and crouching down means that if an accident occurs your face will suffer, whereas your lab coat ought to get the worst of it if you are standing up. The addition of a large globule of liquid adhering to the stopper should be avoided, and the stopper is always placed on the bench upside down to avoid any contamination. After filling to the mark, the flask is stoppered securely, inverted, and shaken (holding the stopper in) thoroughly to mix the contents fully. If the contents are not *immediately* to be transferred to a labelled storage bottle, the outside of the flask is labelled with a descrip tion of the contents. A 'spirit-based' marker pen is useful for this as the ink doesn't rub off on the hands but is easily removed with a tissue (toilet tissue is cheaper than paper hankies!) wetted with methanol or acetone, *etc*.

Pipettes.   Pipettes are designed to transfer a certain volume of liquid from one vessel to another. Various types are used in laboratories, but the most accurate are the so-called 'bulb' pipettes with a single graduation mark [see Figure 5.1(b)]. The same rules for cleaning and maintenance hold for pipettes as for volumetric flasks. Apart from bearing in mind some of the physical and chemical properties of the construction material (usually glass), these are just common sense. The technique for using a pipette is shown in Figure 5.2. The temptation to blow down the pipette to help empty it should be resisted. Pipettes are designed to deliver the appropriate volume whilst leaving a small amount in the tip, and blowing will also introduce saliva into the top.

Even with a skilled operator using a pipette, liquids which give off a potentially harmful vapour are never introduced by mouth suction. These include, for example, concentrated acids, ammonia solution, and organic solvents. A pipette filling device is used (correctly – according to the appropriate manufacturer's instructions), and the best way of learning how to use one is to practise with solutions which will not damage you or the pipette filler if things go wrong!

In general, a significant error in the volume transferred will be incurred with organic solvents because their physical properties (such as vapour pressure and surface tension) are different from those of water, the liquid for which pipettes have been designed.

Burettes.   In order to run a variable amount of a solution into a receiving vessel a burette is used [see Figure 5.1(c)]. The care and maintenance of burettes are as for pipettes and calibrated flasks. For burettes with glass stopcocks the glass surfaces in contact are *lightly* greased. Removing excess grease from inside the tip can be quite tricky, and extremely irritating if the tip blocks during a titration. A length of fine wire can often achieve a temporary clearance, but soaking in a degreasing solvent may be the only long-term solution. A burette should be firstly rinsed carefully with the solution to be used (without putting a thumb over the end) and then filled, if necessary with the aid of a small funnel. The stopcock is opened and the tip filled completely (no air

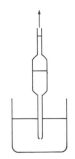

(a) Fill by suction.

(b) Stopper with finger. Wipe excess liquid with tissue.

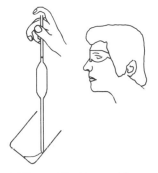

(c) Run out liquid smoothly to graduation mark.

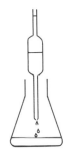

(d) Run into receiving vessel. Touch on sides if possible to avoid splashes and give smooth drainage.

(e) Wait few seconds for drainage and remove (with small volume of liquid). Only touch liquid surface if vessel shape makes wall drainage difficult.

Figure 5.2 *Use of a pipette.*

bubbles), then the burette is filled to just below the initial
graduation mark (there is nothing special about the reading
0.00 cm^3), the funnel is removed, and a few seconds are allowed
for drainage.

Readings of the volume in a burette should be made as shown
in Figure 5.3. After running liquid out, the new reading should
be made only after a few seconds have been allowed for drainage.
Drops of liquid adhering to the walls will cause the reading made
to be larger than it ought and indicate that the walls of the burette
should be degreased.

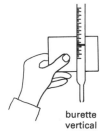

Hold white card or similar
'burette reader' behind burette.
Note reading at particular part
of meniscus. Always measure
at this part of meniscus.

burette   meniscus at eye level
vertical   (use lab stool if necessary)

The reading is between 8.2 and
8.3 cm^3. The second decimal
place would be estimated to be
0.07 cm^3, giving a reading of
8.27 cm^3.

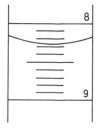

Figure 5.3   *Reading a burette.*

Making a Standard Solution

Weighing will be dealt with shortly, so let's assume that the appropriate weight of a chemical has been weighed into a beaker (and the mass accurately recorded). The operations of dissolution and transfer are shown in Figure 5.4. It should not be forgotten that there may be other components of the solution to be added before making up to volume. For example the chemical may be dissolved in water but the final solution is required in dilute acid. As a general rule, solids are not transferred directly to the calibrated flask, as not all chemicals used for making standard solutions dissolve readily, and crushing and stirring or even heating may be necessary to obtain rapid dissolution.

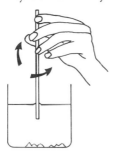

(a) Add some solvent and grind, stir, heat, *etc.* until homogeneous solution obtained.

(b) Put small piece of filter paper between funnel and flask neck to prevent air lock. Pour solution down glass rod.

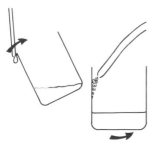

(c) Catch and return last drop, wash down beaker walls, and repeat (b).

(d) Repeat (c) and (b) two or three times.

(e) Make up to volume.

(f) Mix thoroughly.

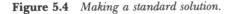

Figure 5.4 *Making a standard solution.*

Performing a Titration

The procedure will involve making a standard solution, which may be the titrant, or more commonly using such a solution to standardise the titrant. The usual procedure is to transfer an aliquot (literally 'an equal part', the requirement being that the volume of solution taken for analysis is an accurately known fraction of the total volume of the solution) of the standard solution by pipette to a titration vessel (usually a conical flask), add any reagents that may be necessary (such as buffer solutions in the case of complexometric titrations) followed by a drop or two of the appropriate indicator, and then add titrant from the burette (having noted the initial reading) until the expected indicator colour change has occurred. The final burette reading is then noted.

To start with, the volume required is unknown, and it is a tedious procedure to carry out an accurate titration since the addition of the titrant must be slow to avoid overshooting the end-point. One way of dealing with this is to carry out a 'rough' titration first. The titrant is allowed to run in rapidly whilst the receiving vessel is swirled to mix the titrant and titrand (the solution being titrated) until the end-point is observed. The volume for this titration will be too high. On subsequent titrations the titrant is run in rapidly until it is known that the end-point will occur within the next 0.50 cm^3 addition. The end-point is approached slowly, drop by drop, the walls of the titration vessel and the tip of the burette being rinsed with water (to ensure that everything run out of the burette has got into the solution) from time to time. Finally, 'split' drop addition is carried out: about half a drop is allowed to suspend from the burette tip and is then washed in with a little water. The accurate titration is repeated at least once.

The process is summarised in Figure 5.5. A 50 cm^3 burette should be read to two decimal places (and not just a zero or five in the last place either). A 10 cm^3 microburette can be read to three decimal places.

Handling Precipitates

The most convenient way of collecting a precipitate in a gravimetric analysis is to transfer it from the precipitating vessel to a sintered glass crucible. Sintered glass is made by heating powder-

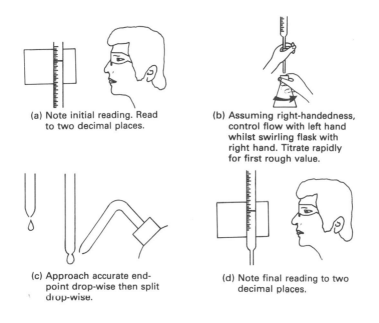

(a) Note initial reading. Read to two decimal places.

(b) Assuming right-handedness, control flow with left hand whilst swirling flask with right hand. Titrate rapidly for first rough value.

(c) Approach accurate end-point drop-wise then split drop-wise.

(d) Note final reading to two decimal places.

Figure 5.5 *Basic titration procedure.*

ed glass until the particles just melt and fuse together. The resulting material is porous, and the sizes of the pores can be controlled by the coarseness of the powder and the extent to which it is heated. A disc of sintered glass about 3.5 cm in diameter forms the base of the filter crucible shown in Figure 5.6. Filtration has to be helped by applying reduced pressure using a 'water pump'. After washing, drying, cooling, and weighing the empty crucible the precipitate is transferred as shown in Figure 5.6. Before transfer the precipitate is allowed to settle, and as much supernatant (the liquid above the precipitate) solution as possible is decanted to avoid clogging the pores of the filter medium early on in the procedure. This is particularly important if filter paper is being used, as a filtration under gravity through a clogged paper can be very slow indeed. Obviously the precipitate should not be 'stirred up' before starting the transfer procedure. A precipitate which has coagulated (fine crystals aggregated together to give large crystals) can be washed in the beaker before transfer. Small portions of the wash liquid are used, the precipitate is swirled and allowed to settle, and the wash liquid is

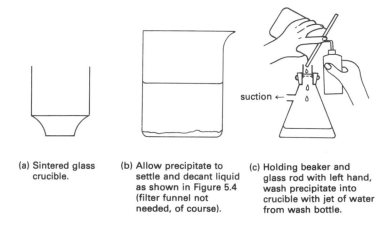

(a) Sintered glass crucible.

(b) Allow precipitate to settle and decant liquid as shown in Figure 5.4 (filter funnel not needed, of course).

(c) Holding beaker and glass rod with left hand, wash precipitate into crucible with jet of water from wash bottle.

Figure 5.6 *Transfer of precipitate into a sintered glass crucible.*

decanted through the crucible. This is more efficient than trying to wash the compacted precipitate in the crucible.

Weighing

Balances used to weigh objects in a chemical laboratory are accurate and precise instruments capable of weighing to the nearest 0.0001 g (0.1 mg or 100 µg) or even lower. For a 100 g object this represents an accuracy of one part in a million. As with all laboratory instruments, balances must be handled carefully and serviced regularly. They should be kept in an environment of stable temperature and free from dust, draught, and mechanical vibration. Weighing on an accurate balance can be speeded up if you know the approximate weight of the object already. This is achieved by weighing on a 'rough' balance (which weighs to the nearest 0.1 g, say). Such rough balances often have a variable-zero control (known as the 'tare'), and in fact the present generation of electronic 'top-loading' balances of both the rough and accurate varieties have this facility. As far as possible all chemicals should be 'weighed by difference'. This not only avoids having to add chemicals to weighing vessels sitting on the pans of accurate balances but also compensates for buoyancy errors arising from air displaced by weighing vessels (but not for those arising from the air displaced by the material being weighed).

Weighing a Chemical to Make a Standard Solution. This is done by 'difference'. As the two weighings on the accurate balance are made within seconds of each other, there is no need to set the zero of the balance to read 0.0000 g. Objects to be weighed (*a*) should not be touched with bare fingers, as a considerable amount of grease and sweat can be transferred (metal tongs, tweezers, or a strip of paper are used), and (*b*) should be at the same temperature as the balance. This is achieved by placing the object in a desiccator in the 'balance room' for a while before weighing. Final equilibration with the balance room atmosphere may be achieved by placing the object inside the balance case. After calculation of the mass of the chemical needed the procedure shown in Figure 5.7 is followed. If quantitative transfer of the solid is required (as,

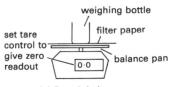

(a) Rough balance.

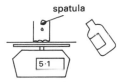

(b) Add approximate mass required. Don't return excess chemical to bottle.

(c) Transfer to accurate balance.

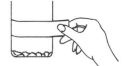

(d) Weigh accurately.

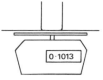

(e) Tip into receiving vessel (for dissolution).

(f) Re-weigh accurately.

Figure 5.7 *Weighing by difference. If the rough balance is 'zeroed' prior to step (a), the approximate mass of the weighing bottle is found. This can save time at step (d) if a mechanical balance is used. If there is any danger of contamination from airborne particulate matter, the weighing bottle should be stoppered.*

for example, when dealing with a solid sample), the contents of the weighing bottle should be dislodged with a fine brush (usually with squirrel hair bristles).

Weighings for Gravimetric Analysis. In these cases the crucible and precipitate will be weighed some time after the initial mass of the empty crucible was found. So before each weighing is made the balance must be set to read 0.0000 g. Similar precautions concerning accidental additions or loss of mass as discussed previously should be taken. Despite their name, desiccators are not particularly good at providing a dry atmosphere, but they do allow a cooling crucible to acquire eventually the same layer of adsorbed moisture as it probably had at the start of the experiment. They also prevent dust particles, clothing fibres, bits of dandruff, *etc.* from settling in the crucible and are thus quite a good place to store crucibles between one analysis and the next.

Hot crucibles (or anything else for that matter) should not be put onto wax-polished wooden benches, and the tongs needed to handle the crucibles should be put down so that the tips don't touch the surface of the bench. In other words, a competent analytical chemist uses common sense and remembers the basic principle of forensic science, namely 'every contact leaves a trace'. The aim in gravimetric work is to keep these 'traces' at levels at which they can't be weighed.

THE UPS AND DOWNS OF PRACTICAL WORK

All chemists should acquire the habit of writing down, in a hard-backed laboratory notebook, weighings, burette readings, and any other relevant practical details necessary for the subsequent calculation. This information should be recorded at the time of conducting the experiment. Some industrial laboratories are required by law to be able to produce such records of the analysis of products on their way to the consumer, and so careful records have to be kept. Sufficient details of the method should be noted down to allow someone else to repeat the experiment, and so should any unusual or unexpected observations. Some teaching experiments may require some 'writing up' afterwards, but in a well designed course this should be minimal, and good chemists will have acquired the habit of 'writing up as they go along'.

Good chemists will also wash up and clean up as they go along, avoiding accidental cross-contamination of solutions and other gross errors. They will even wipe up (spills *etc.*) and wipe down (work surfaces *etc.*) in order to keep their bit of bench as neat and as clean as possible.

HOW RELIABLE IS THE FINAL RESULT?

Quantitative analytical chemistry, like a lot of other physical sciences, consists of making a set of measurements and then performing a set of calculations, involving the numbers arising from these measurements, from which a final number is obtained. This may well be the answer to the original question of 'how much x is there in this sample of y?' In many situations it will be important (perhaps very important) for the person posing the original question to know what reliability can be placed on this value. For example, imagine the following conversation.

Production chemist (PC) in pharmaceutical company: "How much paracetamol was there in that last sample?"

Trainee analytical chemist (TAC): "Ten per cent."

PC: "What do you mean by 'ten per cent'? Ten point nought nought nought? Ten point nought? One times ten to the one?"

TAC: "Er, um..."

PC: "What about the plus-or-minus term? I mean, is it ten point nought nought plus or minus point nought one, or ten plus or minus five?"

TAC: "Ah, yes, I see what you mean."

And I hope you do too. If the production chemist is going to allow tablets, each containing a certain weight of paracetamol, to be made from this material, it is very important that the analytical chemist is able to provide information about the reliability of the result. If the company is going to be prosecuted should the tablets not contain the right amount of paracetamol, it is very important that the analytical chemist can provide the information. In this example it may be just as disastrous to err on the generous side (an expensive business for the company in any case) and put too much drug into each tablet.

So how does the analytical chemist assess the reliability of the final result, and how is this information passed on?

Significant Figures and Decimal Places

Let's start with some basic concepts. When the final result is reported, it is quoted to the appropriate number of significant figures. Significant figures are the digits which are known with certainty together with the first uncertain digit. For example the final reading on a burette might be 23.45 cm^3: the 2, 3, and 4 are known with certainty, but there is some doubt about the 5. It would clearly be meaningless to try and read another digit beyond the five, and so the reading is made to four significant figures. Not all burette readings have four significant figures: for example 9.87 cm^3 and 0.52 cm^3 have three and two significant figures, respectively. All three readings are made to two decimal places, *i.e.* the number of digits to the right of the decimal point is two in each case.

So, if we round off the final result to the right number of significant figures, some information about the reliability of the result is being passed on. Thus 10.00% implies that the required value lies somewhere between 9.995% and 10.005%; 10% implies that the required result lies between 9.5% and 10.5%. In other words, in the absence of any other information, 10.00% means 10.00 ± 0.005% and 10% means 10 ± 0.5%. Rounding off is something that will always have to be done because the number of digits obtained for the final answer will be as many as the calculator will display, and it is ridiculous to quote the result of an analytical procedure to eight or nine apparently significant figures.

But why is it ridiculous to do so? The answer to this involves considering how the numbers of significant figures are affected by calculation. To decide on the number of significant figures in the answer, we have to identify which is the first uncertain digit. This digit together with any digits to the left (apart from zeros, which only serve to locate the decimal point) make up the number of significant figures. When adding or subtracting numbers, it is the number with the least number of decimal places (*i.e.* the greatest *absolute* uncertainty) which governs the overall uncertainty; an example is given in Box 15. When multiplying or dividing, it is the number with the greatest *relative* uncertainty which governs the relative uncertainty of the final result. This rule can be more simply stated as 'the final result cannot have more significant figures than the number in the calculation with the least number of significant figures' (see Box 15).

BOX 15 Significant Figures in Calculations
ADDING OR SUBTRACTING

For example, in calculating a relative molecular mass from relative atomic masses, numbers with different numbers of significant figures will be added. In the case of $Sr_3(PO_4)_2$ the relative molecular mass is $(3 \times 87.62) + (2 \times 30.973\ 762) + (8 \times 15.9994)$, giving $262.86 + 61.947\ 524 + 127.9952 = 452.802\ 724$. As the relative atomic mass of strontium has the greatest absolute uncertainty (± 0.005), the final number must be rounded off to this uncertainty, and so the relative molecular mass is 452.80.

MULTIPLYING OR DIVIDING

For example, when calculating the concentration of a solution after dissolving a known mass in a known volume. Suppose 0.1234 g of material of relative molecular mass 123.45 is dissolved in 10.0 cm^3. The concentration of the solution is $(0.1234/123.45) \times (1/10.0) \times 1000$. The 1000 in the numerator is required to convert the value calculated from mol cm^{-3} to mol dm^{-3}. In this context this is referred to as a 'counting number' and is assumed to have a very large number of significant figures associated with it. As the value of the volume has the greatest relative uncertainty, the final answer must be quoted to three significant figures, *i.e.* 1.00×10^{-1} mol dm^{-3} or 0.0100 mol dm^{-3}.

If you need to be reminded about the rules for rounding off to a certain number of significant figures, look at Box 16.

BOX 16 Rules for Rounding

(1) If the digit to the right of the first uncertain digit is greater (smaller) than 5, increase by 1 (leave unchanged) the first uncertain digit.

(2) If the digit to the right of the first uncertain digit is a 5, and if there are other non-zero digits further to the right, increase the first uncertain digit by 1. Otherwise:
    (*a*) increase the first uncertain digit by 1 if it is odd or
    (*b*) leave it unchanged if it is even.

(3) Round to the first uncertain digit in one step not a succession of steps.

For example, rounding $0.123\ 456$ to three significant figures gives 0.123, rounding $0.123\ 50$ to three significant figures gives 0.124, but rounding $0.122\ 50$ to three significant figures gives 0.122.

Calculations with Numbers Based on Measurements

The situation for a quantitative analysis is a bit more complicated than that outlined above, because the numbers which are used in the calculations have a greater uncertainty than just 1 in the first uncertain digit. This is because most of the numbers have arisen from a physical measurement, which is subject to random errors.

Random Errors

Calibrated Glassware. Supposing you are using a two hundred and fifty cm^3 graduated flask in a method. What volume does it contain? 250, 250.00, 250.000 cm^3? You could calculate the volume by weighing it first empty and then full of water and calculating the mass of water in the flask and hence the volume, from a knowledge of the density of water. To get an accurate answer you might need to correct for buoyancy effects (objects in air are acted on by an upthrust equal to the mass of air displaced – Archimedes' principle). The experiment is quite time-consuming and tells you only about the volume of one particular flask. If you have a cupboard full of them, not to mention all the other sizes (and what about the various sizes of pipettes and burettes?), it is clearly an impossible task to measure the volume of all of them. Instead we rely on the manufacturer to have performed the experiments for us and to tell us what the tolerance is for a particular size of calibrated glassware. This tolerance is expressed as a plus-or-minus term. For 'Grade B' apparatus (the most commonly used grade) some values are given in Table 5.1. The corresponding values for 'Grade A' apparatus are smaller, and of course the price is correspondingly higher. Thus the manufacturer guarantees that at 20 °C your 250 cm^3 flask will contain

Table 5.1 *Tolerances for Grade B volumetric glassware.*

			Flasks (to contain)						
Volume/cm^3	5	10	25	50	100	200	250	500	1000
Tolerance/ ±cm^3	0.03	0.03	0.04	0.06	0.1	0.2	0.2	0.3	0.4

			Pipettes (to deliver)					
Volume/cm^3	1	2	5	10	20	25	50	100
Tolerance/ ±cm^3	0.02	0.02	0.03	0.04	0.05	0.06	0.08	0.12

something between 249.8 and 250.2 cm^3. A 'Grade A' flask would have ± 0.12 cm^3 as the tolerance limits.

The Cumulative Effect of Random Errors.   As random errors are just as likely to be positive as negative, it is unlikely that the overall effect of several random errors will be the maximum possible (*i.e.* the sum of all the individual errors). To account for this, the overall effect is calculated as the square root of the sum of the squares of the relative errors (for terms which are multiplied or divided) or of the absolute errors (for terms which are added or subtracted). The way that errors 'add' together in a procedure is known as the propagation of errors. Consider the following example.

A solution of potassium dichromate is made by dissolving 3.0583 g (the plus-or-minus term associated with weighing an object on an accurate balance is 0.0002 g) in acid and making up to 250 cm^3. This solution is used to analyse an iron(II) solution by titration of 25 cm^3 aliquots. Three accurate titrations give values of 22.11, 22.06, and 22.01 cm^3. What is the concentration of iron (dm^{-3}), to how many significant figures should the result be quoted, and what is the plus-or-minus term associated with the answer?

The equation for the reaction is:

$$6Fe^{2+} + Cr_2O_7^{2-} + 14H_3O^+ \rightleftharpoons 6Fe^{3+} + 2Cr^{3+} + 21H_2O \quad (5.1)$$

The relative molecular mass of potassium dichromate is 294.18, and thus the concentration of the solution prepared is $(3.0583 \times 1000)/(294.18 \times 250.0)$ mol dm^{-3}. There is no need to work this out, so we can go straight to the next part of the calculation. The titration results give a volume of 22.06 ± 0.05 cm^3, and the concentration of iron is therefore:

$$\frac{6 \times 22.06 \times 3.0583 \times 1000}{25.00 \times 294.18 \times 250.0} \times 55.84 \text{ g dm}^{-3}.$$

Using a calculator produces 12.293 883 g dm^{-3}, and applying the simple rule about the number of significant figures (remember the 6 and the 1000 are counting numbers and have lots of significant figures associated with them) gives 12.29 g dm^{-3}. The relative error (the plus-or-minus term), x, is calculated as the square root of the sum of the squares of the relative errors [see equation (5.2)]:

$$\frac{x}{12.29} = \left[\left(\frac{0.05}{22.06}\right)^2 + \left(\frac{0.0002}{3.0583}\right)^2 + \left(\frac{0.005}{55.85}\right)^2 + \left(\frac{0.06}{25.00}\right)^2 + \right.$$
$$\left.\left(\frac{0.005}{294.18}\right)^2 + \left(\frac{0.2}{250.0}\right)^2\right]^{\frac{1}{2}} \tag{5.2}$$

The terms in the numerators on the right-hand side are the ± terms associated with each of the values in the denominator. Unless there is information to the contrary, it is assumed that the uncertainty in the value of a relative atomic or molecular mass is 1 in the first uncertain digit.

$$\frac{x}{12.29} = [(5.14 \times 10^{-6}) + (4.28 \times 10^{-9}) + (8.01 \times 10^{-9}) +$$
$$(5.76 \times 10^{-6}) + (2.89 \times 10^{-10}) + (6.40 \times 10^{-7})]^{\frac{1}{2}} \tag{5.3}$$

$$x = 0.0418 \text{ g dm}^{-3} \tag{5.4}$$

This value confirms that the first uncertain digit is the second decimal place, and thus the value of x is rounded off to 0.04 g dm^{-3} and the result is quoted as 12.29 ± 0.04 g dm^{-3} (with a note to explain how ±0.04 was calculated).

It can be seen that in the expression for x the terms arising from the uncertainties in the relative molecular masses and the weighing are considerably less than those arising from the volumetric manipulations. Omission of the former terms leads to a value of x of 0.0417 g dm^{-3}, which is no different from the value calculated above. Thus, if some terms in the expression are much larger than the others, the others may be omitted and the calculation simplified. The analytical chemist must look critically at the values for any particular analysis and decide whether any terms can be omitted.

Other Sorts of Error

Random errors are not the only sort that the analytical chemist has to consider when deciding on the reliability of a result. There may be errors of the 'gross blunder' category, such as spilling some of the dichromate during transfer from the weighing bottle to the dissolution vessel. This would be serious, as this error would not be detected unless a reference sample (a sample of accurately known composition) was analysed alongside the unknown samples. Other sorts of 'gross blunder', such as not transferring all of the contents of a pipette of sample solution, will show up when the agreement between repetitions of the analysis is examined.

An analytical method may also contain systematic errors. These introduce a bias into the results so that the values obtained are *always* greater or *always* less than the true value. An example of this type of error would be the use of an indicator which indicated an end-point before the equivalence point was reached (see page 91). There may be incomplete precipitation of the desired species in a gravimetric method. Generally speaking, methods which contain systematic errors for the analysis of solutions containing only the analyte species will not be in general use, although the indicator error problem is quite a common one. However, in the analysis of real samples there may well be interference effects from other components of the sample, which could cause the result obtained to be biased. There are also many other sources of systematic error, such as those introduced by the use of instruments or by colour blindness! A good analytical chemist will be constantly on the look-out for such errors.

Accuracy and Precision

A method which contains no systematic errors is said to be accurate, *i.e.* the result obtained reflects the true value for the concentration of the component sought. Methods which do contain these errors are said to be biased or inaccurate. Again, analytical chemists have to use their professional judgement in deciding whether the results from a biased method will be acceptable. In other words, the degree of inaccuracy may have to be assessed.

In order to get the best accuracy from any method, the analysis is repeated several times and the mean (*i.e.* the average) result is calculated.

The uncertainty in the mean result as reflected by the plus-or-minus term arising from the random errors is a measure of the 'precision' of the method. In other words, 'precision' reflects the agreement between successive repetitions of the analysis (known as replicates; two values are referred to as duplicates, three as triplicates). The smaller the ± interval, the higher the precision.

It should be noted that analytical chemists use words like

'precision' and 'precise' in a sense different from their everyday meaning. To the person in the street, 'precise' means 'accurate'.

EVALUATING AN ANALYTICAL METHOD

Analytical chemists must be continually devising new methods or modifying existing ones as new analytical problems appear. Such problems will involve new sample types, new analytes, measurement of lower concentrations, *etc.* In the analytical chemist's toolkit are ways of assessing the performance of new methods and of comparing them with the old methods.

The Background to Precision and Accuracy

A common way of arriving at a value for the ± term associated with the result of an analysis is to repeat the analysis a large number of times and then examine the results. Let's refer to the results as $x_1, x_2, \ldots x_i, \ldots x_n$ (arranged in increasing order of size) and the mean as $\bar{x}$, where $\bar{x} = \frac{1}{n}\Sigma x_i$. A capital Greek 'sigma', Σ, is frequently used to denote 'sum of all things like', in this case like x_i, *i.e.* Σx_i means $x_1 + x_2 + \ldots x_i + \ldots x_n$. If we want to examine the spread of the results, we could calculate the 'range', R, from $R = x_n - x_1$ or we could look at the deviations from the mean, $\bar{x} - x_1$, $\bar{x} - x_2$, *etc.* Some of these numbers would be positive and some negative. To condense this information further, an 'average' deviation is calculated, not just a simple arithmetic mean (which would be zero!), however, but a sort of 'root mean square' deviation. This value is known as the 'standard deviation', s, given by equation (5.5):

$$s = \left[\frac{\Sigma(\bar{x} - x_i)^2}{n - 1} \right]^{1/2} \tag{5.5}$$

The reason for the $n - 1$ in the denominator will be explained shortly. The square of the standard deviation is known as the 'variance'.

The n measurements that have been made are referred to as a 'sample' (a *statistical* sample, that is) taken from the 'background population' of all possible replicate measurements for this analysis. It is clearly impossible to determine experimentally what the background population is, as an infinite number of measurements are required. So what is done is to assume that the background population has a particular form, *i.e.* we use a model for the background. The most commonly used model is called the 'normal' distribution or the 'Gaussian' distribution, which can be represented by equation (5.6):

$$y = \frac{\exp[-(x - \mu)^2/2\sigma^2]}{\sigma\sqrt{2\pi}} \tag{5.6}$$

where y is proportional to the probability of obtaining a particular value of x, μ is the most probable value, and σ is the 'population standard deviation'. The general shape of the curve is shown in Figure 5.8.

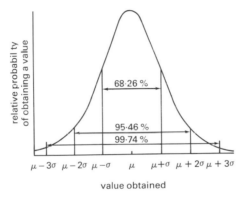

Figure 5.8 *The normal or Gaussian distribution. The model used indicates that the mean is the most probable value, and the numbers indicate the proportion of all values which lie between the 'one sigma', 'two sigma', and 'three sigma' limits.*

In practice, a small number of measurements is made from which we calculate $\bar{x}$ and s (the sample mean and standard deviation); these are used as estimates for the population mean, μ, and standard deviation, σ (these being the true values). The value s, calculated by equation (5.5), is a better estimate of σ than the value calculated by dividing by n. Precision is often numerically expressed as the standard deviation or the relative standard deviation (RSD), given by $s/\bar{x}$, or the percentage RSD, known as the coefficient of variation (CV), given by $100s/\bar{x}$.

The accuracy of the method depends on how good an estimate

x is for μ. Obviously, the smaller the number of replicates, n, and the bigger the standard deviation, the less likely $\bar{x}$ is going to be a good estimate for μ. This likelihood can be expressed by calculating an interval about $\bar{x}$ within which we have a given confidence that μ actually lies. The interval is defined by $\pm ts/n^{\frac{1}{2}}$, where t is a value found from statistical tables. Thus:

$$\mu = \bar{x} \pm ts/n^{\frac{1}{2}} \qquad (5.7)$$

The value of t depends on the level of confidence and the value of n. A confidence level of 95% is often used in analytical work. The wider the confidence interval, the more certain we can be that μ lies within it (thus the 99% confidence interval is wider than the 95% interval) but the bigger is the uncertainty about the actual value of μ. Adopting a 95% confidence level means that, on average, μ will fall outside this range one time in twenty.

Some values of $t/n^{\frac{1}{2}}$ are given in Table 5.2. All analytical chemists should have a calculator capable of calculating s according to equation (5.5). It is then easy to calculate a 95% confidence interval. Thus the 95% confidence interval for the titration value in the example on page 121 is 22.06 ± 0.12 cm^3. Note that this range, 22.18—21.94, is larger than the range of values actually obtained. In order to calculate an overall confidence interval limit for the final calculated result, an overall standard deviation is calculated according to equation (5.2), where numerators are either calculated standard deviations (*e.g.* for titration results) or treated as standard deviations in the case of the manufacturer's tolerance limits, weighing uncertainties, and uncertainties in relative atomic or molecular masses. In this particular example the standard deviation for the titration is 0.05 cm^3 and so the overall standard deviation is 0.0418 g dm^{-3}. The 95% interval is calculated on the basis of the number of titrations as 12.29 ± 0.10 g dm^{-3}; thus the first decimal place is shown to be uncertain and so the result would be quoted as 12.3 ± 0.1 g dm^{-3} (again with a note stating what the $\pm$ term was).

Table 5.2 *Values of t and $t/n^{\frac{1}{2}}$ at the 95% confidence level.*

n	3	4	5	6	7	8	9	10
t	4.30	3.18	2.78	2.57	2.45	2.36	2.31	2.26
$t/n^{\frac{1}{2}}$	2.48	1.59	1.24	1.05	0.926	0.834	0.770	0.715

If the same standard deviation for the titration had been obtained from five titrations, the overall result would be 12.29 ± 0.05 g dm^{-3}. The same standard deviation from ten titrations would give 12.29 ± 0.03 g dm^{-3}. Thus the benefits to be gained from increasing the number of replicates soon begin to fall off. However, a good analytical chemist will always ask the person who wants to make use of the results which $\pm$ term is required. The answer to this question may well influence the choice of methods and the number of replicate analyses.

Outliers

Before starting a calculation it may be suspected that one result is an outlier, *i.e.* it has not come from the same background population as the others, because an undetected error has been made. If a known error has been made, the result can be rejected and the experiment continued, the grounds for rejection being noted in the laboratory notebook. A suspected outlier may be tested by the 'Q-test'. This test is typical of many statistical tests and it goes like this:

(1) we start by assuming that there is no significant effect (the confidence level will have to be specified, *i.e.* how often, on average, we would be prepared to have the test fail),

(2) a value of the required parameter or test 'statistic' according to the appropriate equation is calculated,

(3) we compare the calculated value with a value in a table at the appropriate confidence level, number of observations, *etc.*,

(4) if the calculated value is greater than the tabulated value, the initial assumptions are rejected and we conclude that a significant effect is being observed, or

(5) if the calculated value is less than the tabulated value, we accept the initial assumption and conclude that there is no significant effect.

For the Q-test, Q is calculated from equation (5.8):

$$Q = \frac{x_2 - x_1}{R} \quad \text{or} \quad \frac{x_n - x_{n-1}}{R} \tag{5.8}$$

where x_1 is the lowest result, x_n the highest, and R is the range.

The smallest result is tested first. The test is usually applied to ten results or fewer and the 90% confidence level is chosen. The appropriate values are given in Table 5.3.

Table 5.3   *Values of Q.*

n	3	4	5	6	7	8	9	10
Q	0.94	0.76	0.64	0.56	0.51	0.47	0.44	0.41

With small numbers of results one of them has to be a long way from the others before it will be identified as an outlier. If no grounds for rejection have been observed and the result 'looks' like an outlier, the only thing to do is to continue to carry out replicate analyses until the suspect result fails the Q-test or until so many results have been obtained that a possible outlier does not affect the values of $\bar{x}$ and s very much anyway. When there are more than ten results, a quick test is to calculate the deviation ratio, d, which is the doubtful deviation divided by the standard deviation; if d is greater than 3.3, the result is rejected.

Tests for Comparing Methods

Very often an analytical chemist involved in method development will want to test a new method against an existing one or evaluate two possible methods. Whether one method gives a different result from the other can be assessed by using the 't-test'. Fairly obviously, if the 95% confidence intervals about the two means don't overlap, then the results are different. But how significant is a small overlap? The test statistic 't' is calculated from equation (5.9):

$$t = (\bar{x}_1 - \bar{x}_2)/s(\frac{1}{n_1} + \frac{1}{n_2})^{1/2} \tag{5.9}$$

where s is a 'pooled' standard deviation given by equation (5.10):

$$s = \{[(n_1 - 1)s_1^2 + (n_2 - 1)s_2^2]/(n_1 + n_2 - 2)\}^{1/2} \tag{5.10}$$

The number '$n_1 + n_2 - 2$' is called the 'degrees of freedom', and values of t are looked up for the required confidence level and number of degrees of freedom. For one set of results the number of degrees of freedom is $n - 1$. The t-test can be used to test the accuracy of a method if a reference sample (one whose concentration is known) is available. If information about the reference value (such as mean, number of determinations, and standard deviations) is provided, the full test can be applied to see whether the value being tested is significantly different from the reference value. If such details are not available, a 95% confidence interval can be calculated. If this interval includes the reference value, no significant bias is indicated; if not, the method is shown to be inaccurate.

These equations only apply when the standard deviations of the two methods are not significantly different. You will not be surprised to find that there is a test for this, called the '*F*-test'. A value of F is calculated from $F = s_1^2/s_2^2$, where s_1 and s_2 are allocated so that $F \geqslant 1$. There are lots of other tests, all of which are applied in much the same way.

All quantitative experimental results are useless (and so are values calculated from them) unless they are accompanied by an estimate of the errors involved. Analytical chemists will bear this in mind when (*a*) discussing the possible solution to an analytical problem they are being asked to solve and (*b*) asking someone else to produce analytical results for them.

HOW TO GET MEANINGFUL RESULTS FROM INSTRUMENTS

As you will have gathered, a lot of analytical methods (particularly those for trace analyses of complex mixtures) make use of an instrument. An analytical laboratory will contain quite a few instruments. These will depend on what the laboratory is concerned with analysing but almost certainly will contain balances (though these will probably be in a separate room to prevent corrosion and to isolate them from mechanical vibrations), pH meters, and spectrometers. Many laboratories are involved with analytical methods that require a separation stage and thus will be making use of a number of instrument-based separation techniques (see Chapter 6), and some common laboratory operations (such as titrations) may have been automated to speed up the number of samples that the lab can handle during the working day.

In order to get the best performance out of an instrument, the analytical chemist using it has to know how it works so that it can be set up and operated in a way which does not introduce any unnecessary systematic or random errors into the procedure. The performance of instruments must be continually monitored so that when errors begin to creep in, because a fault is developing or some maintenance is needed, this is spotted right away and the appropriate corrective action is taken.

Generally speaking, instruments have to be 'calibrated' in order to relate the magnitude of what they measure to the amount or concentration of the analyte species. This is because the factors which affect the magnitude of the instrument response to a given concentration are many and complex and often vary considerably from day to day.

In order to calibrate an instrument, standards are prepared. These are solutions of known concentration covering the concentration range likely to be encountered or over which the instrument operates. Sometimes other components of the sample will affect the instrument response (as is common in atomic absorption spectrometry); if their concentration is known, the standards can be matched to the samples. After allowing the instrument to warm up and setting the output to read zero (if appropriate – some instrument scales, such as pH, don't necessarily start at zero, but most spectrometer scales do), the standards are introduced in turn and the instrument response is noted. In addition to the standards, a 'blank' (a solution that contains all the reagents *etc.* but none of the analyte species) is measured. The blank value is not subtracted from the response values for the other standards (it is subject to errors just like the other points).

A calibration curve is obtained by plotting a graph of instrument response on the y-axis and concentration (or related parameter) on the x-axis. The plot is always this way round as the subsequent statistical evaluation usually assumes that the errors in the 'x values' can be neglected, compared with the errors in the 'y values'. This is usually the case with instrumental analytical methods. Often it is expected that there will be a straight-line relationship between the measured parameter and the concentration. In order to calculate this line, the method of least squares is often used (see Box 17). Bearing in mind what has just been said about the usefulness of quantitative data, an analytical result, x_0, obtained by interpolation from a calibration line, should be accompanied by an estimate of the standard deviation given by equation (5.11):

BOX 17 Fitting a Straight Line Through a Set of Points

The basic principle (see Figure 5.9) is to draw the line so as to minimise the deviations of the experimental points from the line in the y direction (the y residuals). As these may be positive or negative, it is the sum of the squares of the residuals that is minimised. Hence the procedure is known as the method of least squares. The general form of the equation of the line is $y = bx + a$, where b is the slope and a the intercept on the y-axis.

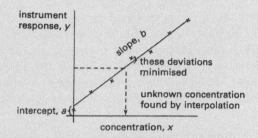

Figure 5.9 *A linear calibration graph.*

The appropriate calculations are most readily performed using a calculator or computer. Values of a and b, their standard deviations, and 95% confidence limits can all be calculated, as can an estimate of how good a fit the points are to a straight line. One such estimate is known as the 'correlation coefficient' and takes values between -1 and $+1$; however, it should be noted that, although a value of ± 1 indicates a perfect straight line of positive and negative slope, respectively, quite scattered points and points which obviously lie on a curve when plotted give apparently quite good correlation coefficients.

If there is any doubt as to the shape of the calibration, the points should be plotted on graph paper; no matter how fancy a computer and data-handling software an instrument contains, if it tries to do something inappropriate with the calibration data, wrong answers will result.

$$s_{x_0} = \frac{s_{y/x}}{b} \left[\frac{1}{m} + \frac{1}{n} + \frac{(y_0 - \bar{y})^2}{b^2 \Sigma (x_i - \bar{x})^2} \right]^{1/2} \qquad (5.11)$$

where:

$$s_{y/x} = \left[\frac{\Sigma (y_i - \hat{y}_i)^2}{n - 2} \right]^{1/2} \qquad (5.12)$$

$\hat{y}_i$ is the value of y on the line for the various x_i values, m is the number of independent replicate measurements of the 'unknown' or 'test' sample, n is the number of calibration points, y_0 is the experimental value from which x_0 is to be found, and $\bar{x}$ and $\bar{y}$ are the mean values of the x_i and y_i calibration values.

The 95% confidence limits may be obtained by multiplying s_{x_0} by the appropriate value of t (for $n - 2$ degrees of freedom). Such confidence intervals can be disappointingly large, which is probably why modern instruments, although they have enough inbuilt computing power to calculate the values, don't usually give them to you!

When interference effects that will cause a certain percentage change in the instrument response are suspected to be present, the analysis may be performed by the method of standard additions (see Box 18). As this is an extrapolative method, the confidence limits are somewhat poorer than the conventional interpolative methods. For both methods the confidence limits can be improved by increasing the number of calibration points (at least six should be taken) and increasing the number of

---

BOX 18   The Standard Additions Method

The basic idea of the method is to ensure that any interference effect present in the sample also acts on the standards. This is achieved by adding increasing amounts of standard to the same amount of sample and diluting the mixtures to the same final volume.

The calibration is a plot of instrument response against concentration added (*i.e.* due to the added standard). The negative intercept on the *x*-axis is the concentration due to the sample (see Figure 5.10). To calculate the concentration in the original sample, this value must be corrected for the sample dilution.

The standard deviation of this intercept is given by equation (5.13):

$$s_{x_E} = \frac{s_{y/x}}{b}\left[\frac{1}{n} + \frac{\bar{y}^2}{b^2\Sigma(x_i - \bar{x})^2}\right]^{\frac{1}{2}} \tag{5.13}$$

and the confidence limits by $x_E \pm t s_{x_E}$.

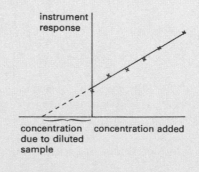

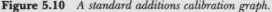

Figure 5.10   *A standard additions calibration graph.*

independent measurements of the unknown y values (which is very tedious, particularly in the case of the standard additions method).

The 'sensitivity' of an analytical technique is given by the slope of the calibration plot (*i.e.* b or some derived value). Despite the everyday usage of 'sensitive', the sensitivity of a method is not a direct measure of how low a concentration can be detected. This 'detection limit' is reached when the instrument response to a low concentration can't be distinguished from its response to the blank because of the random fluctuations (noise) of the response signal.

Although there are several definitions of detection limit in use, one which is readily calculated from a least-squares analysis of calibration data is the x value corresponding to $a + 3s_{y/x}$, namely $3s_{y/x}/b$. Both sensitivity and detection limit are useful parameters for describing instrument performance and thus for allowing a comparison of techniques to be made.

There are some other statistical methods that analytical chemists use when handling calibration data. These can account for a change in precision of each y measurement. This is the usual situation, with the standard deviation increasing as the value of y increases. There are also many situations in which the calibration is curved and a non-linear function has to be fitted to the points. The equations for calculation of confidence limits of an analytical result will be different, of course, but with the ready availability of microcomputers in the analytical laboratory (even in the analytical instrument) there should be no problem in applying the appropriate statistical tests and calculations to the experimental results.

THE ANALYTICAL CHEMIST'S TOOL-KIT

In addition to a range of analytical glassware, the skills required to use it, the ability to transfer 100% of a material from one place to another, and so on, the analytical chemist has to be able to make use of a number of mathematical and statistical tools. To use these tools effectively may require a small computer and appropriate software and will certainly need a good scientific calculator. Portable calculators are the most useful kind, as they can be kept in the lab-coat pocket and are useful devices for recording weighings (sometimes there isn't room to fit a lab notebook beside the balance), especially if the mass is transferred to the calculator's memory (the number is not lost if the device has a battery-saving mode which erases the displayed digits).

The tool-kit will also contain graph paper, ruler, and 'flexi-curve', and the skills necessary to operate correctly a number of laboratory instruments. These will include not only a variety of spectrophotometers but also more basic pieces of equipment such as balances, chart recorders, and pH meters. In addition, the tool-kit will contain a laboratory notebook and the ability to record results and observations as they are made, in a clear, legible fashion, together with the ability to write a concise account of what is being done and to make some sensible comments about the final answers.

Communication skills are important and involve not just the reporting of individual experiments: analytical chemists will have to produce reports of their (and their staff's) activities, discuss problems with other scientists, explain the principles of a method or the function of an instrument to junior staff, outline programmes of work to colleagues, produce work sheets, and so on. Just like other scientists, analytical chemists spend a considerable amount of their time writing and explaining.

Perhaps the most important communication skill of all is the ability to discuss an analytical problem with a non-analyst colleague and explain successfully why an answer to the question that is being asked cannot be provided or why it will come with an uncertainty associated with it.

FURTHER READING

D. Kealey, 'Experiments in Modern Analytical Chemistry', Blackie, London, 1986.

D.C. Harris, 'Quantitative Chemical Analysis', 2nd Edn., W.H. Freeman and Co., New York, 1982.

J.S. Fritz and G.H. Schenk, 'Quantitative Analytical Chemistry', 5th Edn., Allyn and Bacon, Boston, 1987.

H.H. Willard, L.L. Merritt, jun., J.A. Dean, and F.A. Settle, jun., 'Instrumental Methods of Chemical Analysis', 6th Edn., Wadsworth, Belmont, 1981.

J.C. Miller and J.N. Miller, 'Statistics for Analytical Chemists', Ellis Horwood, Chichester, 1984.

R. Schrenfeld, 'The Chemist's English', 2nd Edn., VCH Verlags-gesellschaft, Weinheim, 1986.

C. Turk and J. Kirkman, 'Effective Writing', E. & F.N. Spon, London, 1982.

Chapter 6

Problems with Mixtures – Chemistry to the Rescue

Nearly every analytical method involves either a real or a figurative separation. A real separation means just that: the chemical that we are looking for is separated physically (by precipitation, for example) from the rest of the sample. In a figurative separation the analyte remains in the same place as the rest of the sample but the conditions are adjusted so that only the required species is measured. For example, the conditions may be adjusted so that only the analyte reacts with a certain reagent, as in the use of pH control in a complexometric titration.

Potentially interfering species may be prevented from reacting by masking (*i.e.* the addition of a reagent which forms a stronger complex with the interferent than does the primary analytical reagent but a weaker complex with the desired analyte). The separation may be achieved by using a selective or specific detector, as for example with atomic absorption spectrometry, where a special light source (the hollow cathode lamp) is used whose emitted radiation is absorbed by atoms of only the analyte element. The molecule being sought may be fluorescent whereas the rest of the sample is non-fluorescent, or the analyte may fluoresce in a different part of the spectrum from the fluorescence of other molecules in the sample, and thus 'separation' is achieved by selecting the appropriate wavelength at which to measure the fluorescence. An electrode may be used, responding only to one species in solution.

With the majority of analytical problems, analytical chemists are faced with the task of devising a method that will achieve this 'separation'. This is because most techniques in widespread use are fairly non-specific and most samples contain other components that are likely to interfere. This can be either by giving a false high result, because they undergo, to some extent, the same reactions *etc.* as the analyte, or by giving a false low result, because they prevent the analyte from undergoing the desired reaction or measurement processes. In devising a method to solve such problems, analytical chemists have to use all their knowledge of chemistry, of why things react and why they don't, of how instruments work, of what the interferences are (as explained in the previous chapters), and so on, in order to come up with some sensible suggestions. There is a limit, though, to how far figurative separations can be used. These would be the first choice though, because the method would reduce to: dissolve sample, add reagents, make measurements, and calculate result.

A type of specific reaction chemistry that is increasingly being used is immunoassay (described briefly in Box 19). This is based on the biochemical reaction between the analyte species and antibodies to it, raised in an experimental animal. For suitable analytes the antibody binding reaction can be very specific, and thus the method can be used for analyses of complex multicomponent samples such as blood and urine, with little or no pretreatment. The method usually involves competition between the sample analyte and added analyte labelled in some way. Labelling with a radioactive element is a very popular method at present, but it produces large quantities of low-level radioactive waste with the consequent difficulties of its disposal. A variety of alternative labels are under investigation at present, including fluorescent and chemiluminescent ones (see pages 39 and 41).

Sometimes it is possible to achieve a figurative separation based on the relative speeds of reactions. For example, in the determination of gallium in the presence of aluminium by reaction with lumogallion [4-chloro-6-(2,4-dihydroxyphenylazo)-1-hydroxybenzene-2-sulphonic acid], giving a fluorescent product, up to a ten-fold excess of aluminium can be tolerated because the reaction of gallium with the reagent is about 1000 times faster than the reaction of aluminium. Such 'kinetic masking' (as this differential reaction rate method is called) can also be achieved by the use of catalysts. As with reaction chemistry, the most specific types of catalysts are the biochemical variety, *i.e.* enzymes.

For example, glucose can be determined in urine (an important

BOX 19   Principles of Immunoassay

When a sufficiently large molecule (relative molecular mass of 6000 or more) is injected into an animal, it triggers the animal's immune response system, which manufactures antibodies to the foreign molecules (called antigens). The antibodies are γ-globulins (proteins) which bind specifically to the antigens and to nothing else. Antibodies to smaller molecules can be raised, but such molecules usually have to be bound to a larger protein molecule first. After a few weeks some blood is taken from the animal and the serum, which contains the antibodies, is separated. No further purification is usually necessary and the reagent is used in this form (called antiserum).

The general scheme of an immunoassay is to mix a known amount of *labelled* antigen ($Ag^{\star}$) with the antigen (Ag) containing the sample and to add more antiserum (diluted to contain fewer antibodies, Ab, than are required to react completely with Ag and $Ag^{\star}$). Provided that the label does not affect the antibody's ability to recognise the antigen, competition for the binding sites on the antibody occurs, and some labelled and unlabelled antigen will remain unreacted. This may be represented as a 'before and after' situation [reaction (6.1)]:

$$Ag^{\star} + Ag + Ab \; \rightleftharpoons \; Ag^{\star}Ab + AgAb + Ag^{\star} + Ag \qquad (6.1)$$

before reaction   after reaction

The amount of $Ag^{\star}$ is constant and known, so, as the amount of Ag increases, less $Ag^{\star}Ab$ will be formed and a greater proportion of $Ag^{\star}$ will remain unbound after the reaction. The final stage of the assay involves measuring the amount of either $Ag^{\star}Ab$ or $Ag^{\star}$ (or both). Unless the label has changed in some way that can be detected when bound to Ab (and this is possible with certain labels), $Ag^{\star}Ab$ and $Ag^{\star}$ have to be separated first. There are a variety of methods of achieving this, including reaction of $Ag^{\star}Ab$ with a second antibody which is 'anti' the first antibody, with precipitation of the resulting complex. The design of immunoassay procedures presents a considerable challenge to analytical and synthetic chemists, and a wide variety of different labels (see pages 41 and 52) and separation steps (where necessary) have been devised; the area is being actively researched at present. One interesting recent development is in the use of antibodies bound to a solid support material, such as plastic beads. Separation of the bound and unbound antigen may be achieved simply by washing the unbound antigen off the beads.

analysis in the diagnosis of diabetes and subsequent monitoring of diabetics) using the enzyme glucose oxidase, which specifically speeds up the reaction glucose + O_2 + $H_2O \rightleftharpoons$ gluconic acid + H_2O_2. The hydrogen peroxide can also be measured enzymatically using another enzyme, peroxidase, to catalyse the reaction H_2O_2 + I^- + $2H^+ \rightleftharpoons I_2$ + $2H_2O$. The decrease in I^- concentration is followed by using an iodide-sensitive electrode (see Chapter 3 for general principles), although spectrophotometric methods are more commonly used in clinical laboratories.

There is a limit, though, to the amount of selectivity that can be achieved with chemical reactions. The problem is that, as the sample becomes more and more complicated, with more and more components similar to the desired analyte, sufficient differences in ΔG° (and hence K) or rates for the reactions of analyte and of interferents (so that the analyte has reacted completely but interferents have not) cannot be obtained. Also, many analytical problems involve the determination not just of a single component in the sample but also of several (possibly similar) or even all components.

REAL SEPARATIONS

These involve a physical redistribution of the sample components so that the component of interest becomes located at a different place from the remainder of the sample. To start with, let's just consider separations that divide the sample into two parts: the component, or components, of interest and the remainder of the sample.

Phase Change

It may be possible to precipitate the analyte (or 'co-precipitate' it – in this technique a small amount of the desired material is brought out of solution by a large, bulky precipitate). It may be possible to separate the analyte from the remainder of the sample by distillation or crystallisation. It is possible to use solid reagents with reactive groups on their surface (normally referred to as immobilised reagents). For example, there are resins with chelating groups on their surface (they've been chemically synthesised, of course) a bit like parts of the edta molecule (see page 73) that will bind metals.

There are also resins with positively or negatively charged groups on their surface that will hold negatively or positively charged species, respectively. Such resins are often used in a preconcentration procedure whereby a large volume of a dilute sample solution is passed through a bed of resin. All the analyte is retained and separated from the sample components into a fairly small volume and is then stripped from the solid reagent by a displacement reaction, such as washing with concentrated acid or alkali. For example, several trace metals (such as lead and cadmium) can be determined in sea-water by preconcentration on a chelating resin followed by dissolution into a small volume of concentrated acid and determination by flame atomic absorption spectrometry. The different oxidation states of iron can be measured by acidifying the sample with hydrochloric acid and passing the solution over a positively charged resin which retains $FeCl_4^-$ but allows Fe^{2+} to pass on and be measured. The $FeCl_4^-$ can be removed by washing with water, when $FeCl_4^-$ dissociates to give Fe^{3+}, which is not retained by the resin.

Size

The use of a bed of resin to retain selectively certain components of a sample solution being passed through is similar in concept to filtration. The familiar medium of filter paper separates solid material, too big to pass through the pores of the paper, from the liquid. This process can be performed at the molecular level using a specially made polymeric gel as filter medium with pore sizes that can discriminate between molecules (or other species in solution) of different sizes.

A variety of gels are used, and a common one is based on a crosslinked linear glucose polymer called dextran. The technique is called gel filtration. It is a useful method for cleaning up and preconcentrating a sample in which it is the large molecules that are of interest. Passage through a short gel column will produce a solution of large molecules, excluded from the gel pores, in the small void volume (*i.e.* the volume of liquid between the gel beads). Alternatively, the sample solution may be added to the dry gel, which swells and takes up the small molecules, including water, leaving a concentrated solution of the large molecules.

Another method based on size discrimination is known as dialysis. With this technique the sample is contained entirely within a membrane (a bit like a sausage skin) of a particular pore size. Cellophane, for example, has a pore size of approximately 6 μm and blocks molecules with a relative molecular mass of greater than 10 000. The dialysis membrane containing the sample would be left suspended in a solution of low concentration of the molecules or ions to be removed, the driving force for the separation being the concentration gradient between the inside and the outside of the membrane. Diffusion across the membrane 'ceases' when the concentrations either side are equal.

Solubility

Although there are rather limited possibilities for dealing with mixtures by precipitating various components with appropriate reagents, the extraction of the desired component into another solvent system forms the basis of a useful separation method. Generally speaking, some chemistry has to be performed on the species of interest in order to make it soluble in the second solvent. The analytical chemist can assess the possibility of achieving a separation by this method of liquid–liquid extraction (or 'solvent extraction', as it is sometimes called) by considering the factors which govern the solubility. The most commonly used system is to extract the species of interest from an aqueous system into an immiscible (or very slightly soluble) organic solvent. In general, uncharged, large, bulky, non-polar molecules will be more soluble in organic solvents (it takes less energy to form a big enough 'hole' in the organic solvent than it does to form the same hole in water). Therefore essentially covalent, neutral molecules, uncharged metal chelates, and ion association complexes will be extracted.

The situation is summarised in Box 20 and Figure 6.1. It should be noted that there are various equilibria operating and that control of the selectivity can be achieved by pH control and masking in a manner similar to that discussed for complexometric titrations (see pages 98 and 99).

Analytical problems which centre around how to make the overall procedure specific for one particular component can often be tackled successfully by the techniques described so far, most of which can both separate and preconcentrate the desired component – a useful feature for trace analytical problems. The methods are less useful for the multi-component type of prob-

BOX 20   Liquid–Liquid Extraction

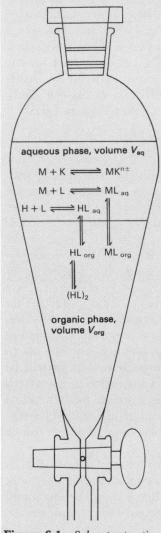

aqueous phase, volume V_{aq}

$M + K \rightleftharpoons MK^{n\pm}$

$M + L \rightleftharpoons ML_{aq}$

$H + L \rightleftharpoons HL_{aq}$

HL$_{org}$ ML$_{org}$

(HL)$_2$

organic phase,
volume V_{org}

The extraction of a neutral species may be improved if it dimerises in the organic phase (*e.g.* carboxylic acids in tetrachloromethane), reducing the polarity of the species in this phase. The extraction of a metal, M, may be prevented if it forms a stronger charged complex with a ligand, K, than it does a neutral complex with the extractant, L. The extraction will also be affected by the extent of solvation of the extractable species in both aqueous and organic solvents.

The ratio of the concentrations of the extracted species in each phase is known as the partition coefficient, K_D.

Thus $K_D = [ML]_{org}/[ML]_{aq}$ for ML species. The distribution ratio, D, is given by $D = [ML]_{org}/([ML]_{aq} + [M]_{aq})$, *i.e.* the ratio of the total concentration of solute in each phase. This value of D is for the M species in the absence of any interfering ligand. The percentage extraction is given by equation (6.2):

$$E(\%) = \frac{100D}{D + (V_{aq}/V_{org})} \quad (6.2)$$

It is easy to calculate from this equation that it is more efficient to extract with several small volumes of organic solvent than one large volume (of the same size as the sum of the small volumes).

Figure 6.1   *Solvent extraction.*

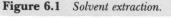

lem, which really requires the separation of various analyte components from each other as well as from potentially interfering components of the sample. Such separations are achieved by employing the appropriate type of chromatography. Chromatography is one of the most powerful and widely used separation methods. In particular, instrumental chromatography (in which separation and measurement are both carried out in one instrument) has made tremendous strides in the last few years, and some of the country's best scientists have been (and are still, in some cases!) very active in the development of the various techniques. The contribution that the techniques of chromatography have made to virtually every area of science cannot be overstated, and the originators of modern instrumental chromatography, A.J.P. Martin and R.L.M. Synge, were rightly awarded their Nobel prizes.

Chromatography Gets One Phase Moving

The basic principle of chromatography is to get the components of the sample in contact with two phases and to move one phase relative to the other. The components of the sample distribute themselves between the phases according to their relative affinities, and separation is achieved between components which have different relative affinities. Components which do not interact with the stationary phase move with the speed of the mobile phase, and components which interact strongly with the stationary phase don't move at all. These two extremes of behaviour are, in fact, the immobilised reagent method of separation already described. Between these two extremes other components will move with speeds governed by how long they spend in the mobile phase and how long in the stationary phase. These times are governed by partition coefficients (see Box 20), and thus the speed of the various components depends on their partition coefficients and a separation will be achieved when partition coefficients differ. A widely used type of chromatography has liquids as both phases.

Imagine an experiment in which we perform a solvent extraction from phase S to phase M (shake to equilibrate, allow layers to separate), then transfer phase M to another separating funnel which already contains some of phase S, and add some more phase M to the first funnel. Equilibrium is now achieved in both funnels and the process repeated so that phase M in funnel 2 is transferred to funnel 3 (which already contains phase S), the phase M in funnel 1 is transferred to funnel 2, and fresh phase M is added to funnel 1. And so the process goes on, as shown in Figure 6.2.

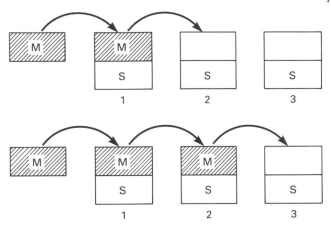

Figure 6.2 *Solvent extraction with one phase on the move. The upper picture is for after the first extraction, the lower for after the second.*

What is going to happen to a solute, A, which is distributed between the 'mobile' phase, M, and the stationary phase, S? It doesn't take much imagination to realise that A is going to be distributed between all the funnels that the leading portion of phase M has reached and between each of the two phases in the funnels. The particular nature of the distribution between phases and funnels will depend on K, the distribution or partition coefficient, and the volumes of phases S and M in the funnels. Let us take the very simple case in which the volumes are equal and there are no competing side reactions, so that $K = D = [A]_m/[A]_s$ (see Box 20). It's quite straightforward, but rather tedious, to work out the fraction of the original amount of A in each funnel. It turns out that for the nth funnel the fraction of A is $\left(\dfrac{1}{K+1} + \dfrac{K}{K+1} \right)^n$.

It is very interesting indeed to present the results graphically for different values of K. This is done in Figure 6.3 for $K = 0.2$, 1.0, and 5.0. It can be seen that, if the original solution contained two solutes which had values of K of 0.2 and 5.0, these two components would have been completely separated in different funnels and could be recovered by collecting and pooling the contents of the appropriate separating funnels. A component with K of 1.0 would not be completely separated from either of the other two components, but there is only a small amount of

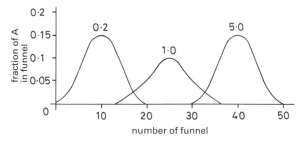

Figure 6.3 *Distribution of solute between funnels after 50 extractions for values of the partition coefficient, K, of 0.2, 1.0, and 5.0.*

overlap and the pure component could be obtained from any number of funnels between about numbers 18 and 32 if the material had to be investigated qualitatively (*i.e.* to find out what it is).

If the experiment was continued and the portions of phase M (the mobile phase) were collected as they were transferred out of funnel 50, then eventually all of the components would emerge from the 'system' and, if we could detect the presence of the components in the mobile phase, it would be possible to observe when one component has been collected and thus to collect the components in separate receiving vessels. You can also imagine that the more funnels there are, the better the separating power of the system, and so the components with more similar *K* values could still be separated. However, the more funnels there are, the more any given component is distributed between them, so, if the situation illustrated for three components in Figure 6.3 was shown for 100 funnels, the 'peaks' would be broader.

In a chromatography system the boundaries between the successive funnels are 'removed' (see Figure 6.4). This is achieved by immobilising the stationary phase S on some suitable solid

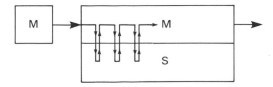

Figure 6.4 *The basic chromatographic experiment. The small arrows indicate the start of the journey taken by a hypothetical solute molecule.*

support and arranging some means by which the other phase can be made mobile.

Paper, Thin Layers, and Columns. The cellulose fibres of ordinary filter paper hold water, which will act as the stationary phase. The mobile phase can be induced to move simply by dipping one end of the paper into a tank of solvent, whereupon it moves through the paper by capillary action. The paper can be held horizontally or vertically and the mobile phase can move up or down (gravity assisted). The way the technique is used is illustrated in Box 21 and Figure 6.5. For example, amino acids can be separated using 1-butanol : ethanoic acid : water (4:1:5 by volume) as the mobile phase. The separated amino acids may be visualised by spraying with indane-1,2,3-trione (commonly referred to as ninhydrin), which produces a purple coloration. This type of separation, involving two liquid phases, is known as partition chromatography.

Chromatography can also be carried out with a finely divided solid (rather than the fibrous structure of paper) on a thin layer or in a column. In these cases there may not be a stationary liquid phase, but the 'partition' effect that acts as the basic separation mechanism may be between 'active' sites on the solid support (highly polar or charged groups) and the mobile phase. The case of highly polar groups (*e.g.* silica gel or alumina) is known as adsorption chromatography; if charged groups are involved, the process is ion exchange chromatography. In practice, the detailed mechanism of the separation for any given stationary material and mobile phase may involve partition, adsorption, and ion exchange.

Thin-layer chromatography (usually referred to as TLC) is a very widely used technique for checking the purity of materials. If only one spot is obtained in a series of experiments using a variety of solvents of different polarity, then it is likely that the material is pure. It is simple, rapid, and relatively cheap and gives better separations than paper; however, like paper chromatography, is difficult to apply quantitatively. It can be made very versatile as far as spot visualisation is concerned, by coating the layer with a fluorescent material whose fluorescence is quenched by the presence of a 'spot'. After development and drying, the plate is examined under ultraviolet light.

All three types of chromatography can be performed with the stationary phase packed into a column and using gravity to drive

BOX 21   Paper Chromatography

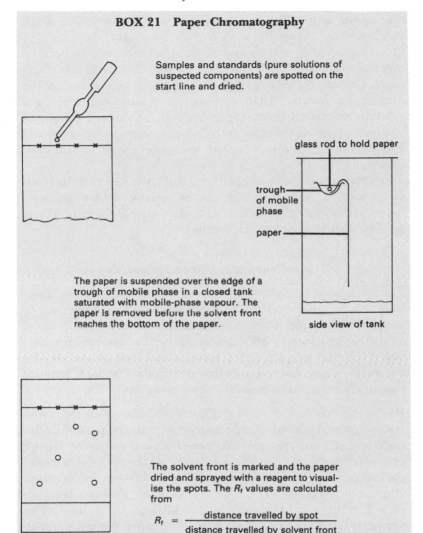

Samples and standards (pure solutions of suspected components) are spotted on the start line and dried.

glass rod to hold paper

trough of mobile phase

paper

The paper is suspended over the edge of a trough of mobile phase in a closed tank saturated with mobile-phase vapour. The paper is removed before the solvent front reaches the bottom of the paper.

side view of tank

The solvent front is marked and the paper dried and sprayed with a reagent to visualise the spots. The R_f values are calculated from

$$R_f = \frac{\text{distance travelled by spot}}{\text{distance travelled by solvent front}}$$

Figure 6.5 *Paper chromatography.*

Components are identified from agreement between R_f values. Confirmation is obtained by running another chromatogram using a different solvent system. Sometimes the solvent front is allowed to run off the bottom of the paper and R_X values are calculated in an analogous fashion to R_f values, but in this case the denominator is the distance travelled by substance X, usually a standard.

the mobile phase. With a column it is usual to collect small volumes of the eluent (the mobile phase coming off the bottom of the column) and test them for the presence of a separated component. These small collected volumes are known as 'fractions'. Column chromatography is frequently used as part of the organic chemist's tool-kit to separate the various components of a reaction mixture. Chromatographic separation on the basis of size is also possible using a controlled-pore-size stationary phase. This technique is known as size exclusion or gel permeation chromatography.

Despite the very wide range of types of chemical, ranging from metal ions to proteins, which can be separated by these basic forms of chromatographic methods, they are not the most widely applied analytical chromatographies.

The Instrumental Chromatographies

The most widely applied chromatographic methods are those based on the separation and quantitative detection of the components of the sample. There are two basic types: the group based on the use of a liquid mobile phase (known as high-performance liquid chromatography, HPLC) and those based on the use of a gas mobile phase and usually a liquid stationary phase (therefore known commonly as gas–liquid chromatography, GLC).

Putting the HP into LC. The 'high performance' is the rapid separation and quantitation of many components, possibly at the trace level, in mm^3 sample volumes. This is obtained by having very small particles (down to a few μm) of the support material packed uniformly into a column (typically 10—25 cm long and 4 mm internal diameter). To get a reasonable flow rate through such a column a high-pressure (20 kPa) pump is used. The sample is injected as a small volume at the top of the column and, after separation on the column, the eluent passes through a flow-through detector which continuously monitors the concentration of the separated components. A schematic diagram of an HPLC instrument is shown in Figure 6.6. The resulting readout is called a chromatogram and is a plot of detector response against time.

The requirement to have a constant flow of mobile phase of a few cm^3 per minute places quite a demand on the quality of engineering and plumbing. The pump must be capable of delivering a pulse-free flow and be inert towards the various

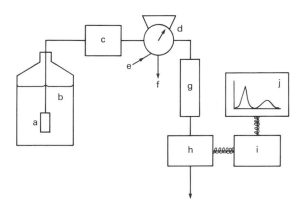

Figure 6.6 *Schematic diagram of an HPLC apparatus:* a, *filter;* b, *mobile phase;* c, *pump;* d, *injection valve with sample loop;* e, *sample inlet;* f, *waste;* g, *column;* h, *detector;* i, *detector signal processor;* j, *readout.*

solvents used. All the tubing is narrow-bore stainless steel, and the column and end fittings are all manufactured from this material. As well as a smooth, uniform flow, precise results will only be obtained with precise injections. So a high-quality injection valve capable of operating at high pressures is required. The volume injected is only a few mm^3. Great care has to be taken not to block any of the connecting tubing or the column inlet with dust particles, and so the mobile phase is filtered on its uptake from the solvent reservoir. The sharpness of the zones that the components of the sample have when they are eluted from the column must not be distorted by the detector. If the detector had a big internal volume, it would mix up the separated zones and all the separation produced by the column would be lost. Thus detector design is quite a tricky problem.

Various analytical techniques have been adapted to the detection of material in LC eluents, including UV and visible absorption and fluorescence spectroscopies, voltammetry, and measurement of refractive index. The latter phenomenon is not particularly sensitive but is applicable to a wide range of species, whereas the other types of detector do not respond to all types of molecules.

The extent of the separation achievable depends on a number of operating parameters of the system, including the nature of stationary phase, nature of mobile phase, temperature, and flow rate. The length and diameter of the column and the volume of sample injected can also be changed, so the chromatographer (as the analytical chemist who specialises in this aspect of separation science is known) has a wide variety of experimental variables to ring the changes on. Some separations can be improved by changing the nature of the mobile phase during separation. This is done by having two, or even three, solvent reservoirs with a control mechanism to change gradually the composition of the mobile phase as the separation is proceeding. As with all types of chromatography, the factors which will lead to a good separation also lead to long analysis times, so the analytical chemist has to be able to optimise the separation by suitable choice of variables in order to achieve acceptable separation of components in the minimum time. It is thus important that the basic principles of the separation mechanism be understood.

Measuring Separation.   Good separation between components is not just a matter of getting the maxima of adjacent peaks as far as possible; it also depends on how broad the peaks are. The parameters used to measure the separation and to describe peaks in chromatograms are shown in Box 22. At resolution $R = 1.0$ adjacent peaks are 98% separated; at $R = 1.5$ they are 99.5% separated. The parameters governing resolution can, to some extent, be independently varied by changing the mobile phase, stationary phase, temperature, and ratio of the amount of mobile phase to stationary phase. The column efficiency, N, can be changed by changing the length of the column. N is sometimes known as the 'number of theoretical plates'. This terminology is a hangover from a time when fractional distillation terminology was applied to chromatographic behaviour. The value of N is often used as a rough guide to the 'quality' of a particular chromatographic system. The length of the column, L, divided by N is known as the HETP (height equivalent to a theoretical plate), *i.e.* the length of column equivalent to one plate. It is used to compare columns of different lengths. Box 22 shows that the separation between the peak maxima is governed by the thermodynamics of the partitioning process for the two components involved. But what causes peak broadening? The various factors are illustrated in Figure 6.9. They can be expressed mathematically as an equation which relates HETP to the flow rate dependence of these parameters, which can be useful in predict-

BOX 22　Quantitative Parameters in Chromatography

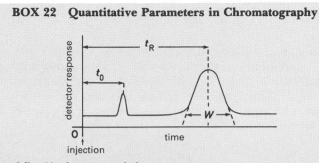

Figure 6.7 *Single-compound chromatogram.*

Figure 6.7 shows the chromatogram of a solution of a single compound. The first small peak corresponds to the unretained solvent. This does not interact with the stationary phase (not generally true in practice) and travels through the column with the mobile phase, and t_0 is thus the time the mobile phase takes to get through the column. If the volumetric flow rate is f_v, the volume of mobile phase in the column is $V_0 = t_0 f_v$. The compound is retained in the column because it interacts with the stationary phase to a certain extent, and the peak maximum appears at a time t_R, known as the retention time. The volume of eluent which has flowed to elute the peak maximum is $V_R = t_R f_v$. The number of column volumes of mobile phase (*i.e.* the number of V_ms) required to elute the peak maximum over and above that required to elute an unretained compound is known as the capacity factor, k', given by equation (6.3):

$$k' = \frac{V_R - V_m}{V_m} = \frac{t_R - t_0}{t_0} \tag{6.3}$$

This parameter has been called by a variety of names in the past and is now usually referred to as simply 'kay dashed'. The quantity $t_R - t_0$ is known as the adjusted retention time, t_R'.

Now the fraction of its time in the column that a molecule spends in the stationary phase is equal to the fraction of the molecules in the stationary phase at any instant, and the fraction of its time in the mobile phase is equal to the fraction of the molecules in the mobile phase at any instant. Taking the ratio of these fractions gives equation (6.4):

$$\frac{t_R - t_0}{t_R} \cdot \frac{t_R}{t_0} = \frac{n_s}{n_s + n_m} \cdot \frac{n_s + n_m}{n_m} \tag{6.4}$$

where n_s and n_m are the number of molecules in the stationary and mobile phases, respectively, at any instant. If V_s is the volume of the stationary phase then:

$$\frac{n_s}{n_m} = \frac{c_s V_s}{c_m V_m} \tag{6.5}$$

where c_s and c_m are the concentrations in the stationary and mobile phases, respectively. In general, the ratio of the concentration in the stationary phase to that in the mobile phase, c_s/c_m, is called the partition coefficient, K. (This is the commonly used notation despite the possible confusion with terms from liquid–liquid extractions – see Box 20.) Therefore:

$$\frac{t_R - t_0}{t_0} = \frac{KV_s}{V_m} \tag{6.6}$$

and from equation (6.3):

$$\frac{KV_s}{V_m} = k' \tag{6.7}$$

Thus k' is the ratio of the amount of solute in the stationary phase to the amount of solute in the mobile phase (as well as the parameter with the long-winded definition given previously). Equation (6.7) is a fundamental relationship in chromatography.

Peaks are assumed to be Gaussian in shape (predicted by considering the random nature of the processes contributing to band-broadening), and thus the base width, W (obtained by drawing tangents at $t_R \pm \sigma$), is 4σ, where σ is the standard deviation of the peak (see page 125).

The column efficiency, N, is given by equation (6.8):

$$N = \left(\frac{t_R}{\sigma}\right)^2 \tag{6.8}$$

or by $16\left(\dfrac{t_R}{W}\right)^2$ or $5.54\left(\dfrac{t_R}{W_{1/2}}\right)^2$, where $W_{1/2}$ is the width at half the maximum height (the width of a Gaussian peak being 2.38σ). The resolution between the two peaks in Figure 6.8, R, is given by equation (6.9):

$$R = \frac{2(t_{R2} - t_{R1})}{W_2 + W_1} \tag{6.9}$$

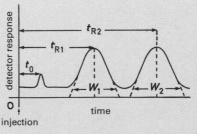

Figure 6.8 *Separation of two components.*

ing the effect of changing the flow rate on the band spreading. The effect of this parameter on analysis time must not be overlooked.

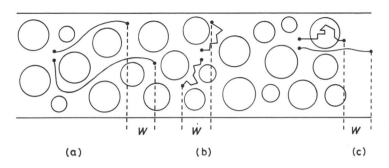

Figure 6.9 *The major factors contributing to band-broadening:* (a) *tortuous nature of flow paths (sometimes called eddy diffusion),* (b) *diffusion along the column,* (c) *slow equilibration between mobile and stationary zones (sometimes called resistance to mass transfer). The spacing W is the contribution to the overall band width due to the particular factor. All these factors are kinetic in nature, whereas separation between components is governed by thermodynamic factors.*
(Reproduced with permission from 'High Performance Liquid Chromatography', ed. J.H. Knox, Edinburgh University Press, 1978.)

Nature of the Stationary Phase. An important development early on in HPLC was the use of 'bonded phases'. These are materials in which the stationary phase is covalently bonded to the support. Most bonded phases are based on silica gels, and, although polar groups such as aminopropyl, cyanopropyl, ether, and glycol are available, it is the non-polar bonded phases that are widely being used. Various hydrocarbon chain lengths are available, but the most popular is the octadecyl group ($C_{18}H_{37}$), and the material is referred to as ODS (octadecyl silyl) silica. Chromatography with this material has the more polar phase as the mobile phase and has thus come to be known as reversed-phase chromatography. The stationary phase bonded to the support material is like a sort of hydrocarbon 'fur' which acts as an organic solvent. Reversed-phase chromatography is primarily useful for separating polar molecules, but it will separate non-polar molecules as well. Care has to be taken that the mobile phases used are not too basic or acidic, as the stationary phase can be cleaved off the support by hydrolysis. However, new developments in this area are continually being reported, and commercially available ODS phases are increasingly stable. In addition to partition chromatography, bonded phases are available for ion exchange and for gel filtration (size exclusion) chromatographies as well. Suitable material is also available for reversed-phase paper and thin-layer chromatographies.

Analytical Applications. HPLC is usefully applied to the separation of small molecules and ionic species of medical or biological origin and of a variety of compounds of high relative molecular mass or low stability such as proteins, nucleic acids, amino acids, sugars, polysaccharides, lipids, synthetic polymers, surfactants, dyes, and pharmaceuticals. The last category is the main area of application, and as an example Figure 6.10 shows the analysis of a mixture of sulphonamides (drugs with antibacterial activity). Detection was by UV absorption at 254 nm. Notice how quantitative analysis of the first two components is only possible when gradient elution is applied to produce good resolution. Confirmation of the identity of an unknown can be obtained by adding the pure component and observing the change in peak height. To make absolutely sure, this experiment should be repeated with another chromatographic system. The peak height and area are both related to the amount of material present via the relative response of the detector for that particular component. Quantitative analysis can be done by first establishing the response of the components relative to that for a similar compound, added in known amount as an internal standard. Alternatively, the system can be calibrated for each component by injecting known amounts.

Gas–Liquid Chromatography

This is the other very widely used instrumental chromatography. Until the comparatively recent emergence of HPLC techniques, GLC was the only chromatographic technique that gave high performance in terms of resolution and detecting ability. But, of course, the fact that the sample has to be chromatographed as a vapour places some restrictions on the properties of the molecules that can be handled by this technique. Roughly 15% of all organic compounds are sufficiently volatile and thermally stable to be handled by GLC. One way of getting round this limitation is

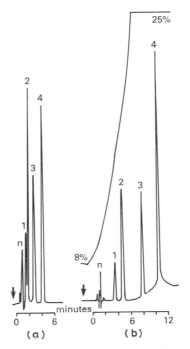

Figure 6.10 *The HPLC analysis of a mixture of sulpha drugs (sulphon-amides): (1) sulphadiazine, (2) sulphamerazine, (3) sulpha-methoxazole, (4) sulphaquinoxaline. An ODS stationary phase was used. Chromatogram (a) was obtained with isocratic elution (no change in mobile phase) with 25:75 (v/v) acetonitrile : wa-ter; (b) was obtained with gradient elution as shown from 8:92 to 25:75 (v/v) acetonitrile : water. n denotes a non-retained solute.*
(Reproduced with permission from 'High Performance Liquid Chroma-tography', ed. J.H. Knox, Edinburgh University Press, 1978.)

to convert the sample components to more volatile derivatives; for example, with sugars the conversion of ROH to $ROSiMe_3$ (the trimethylsilyl ether) is readily achieved with commercially avail-able reagents. The derivative is thermally stable and volatile. Although the advent of HPLC has reduced the requirement for the production of volatile derivatives, GLC is still a useful technique for the sorts of molecules that are difficult to detect by HPLC, such as alkanes or alcohols, which don't absorb UV or visible radiation. Another well established technique for produc-ing volatile material from the sample is to pyrolyse it (heat rapidly in the absence of air) and thus break it down into characteristic fragments. This is widely used for the study of polymers in the analysis of paints, plastics, and rubbers.

A schematic diagram of the components of a gas–liquid chromatograph is shown in Figure 6.11. The carrier gas is

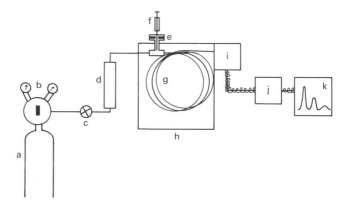

Figure 6.11 *Basic gas chromatograph:* a, *high-pressure carrier gas reservoir;* b, *pressure regulator;* c, *valve;* d, *flow meter;* e, *injection port;* f, *sample injection from syringe;* g, *chromatographic column;* h, *oven;* i, *detector;* j, *amplifier and signal processor;* k, *readout.*

commonly nitrogen flowing at 30—40 cm^3 min^{-1} through a chromatographic column up to 3 m long, 3—4 mm internal diameter (coiled for convenience), which is placed in an oven. The sample is introduced by injection of a few mm^3 of solute in the vapour phase through a self-sealing rubber septum and is flash-evaporated onto the column. The 'liquid' stationary phase is normally a high-molecular-weight grease or oil of low volatility coated onto small-diameter inert support particles. The loading is maybe only 3—10% and the column packing material is a free-running powder at room temperature, and columns can be packed relatively easily. Some stationary phases, such as the phases based on silicone oil, can be operated at temperatures of up to 350 °C.

The quantitative bases of the separations are exactly as discussed in Box 22 for liquid chromatography, and exactly the same equations apply, as do considerations of how to improve the analytical performance for a given problem. The broadening factors, outlined in Figure 6.9, also apply to GLC. However, there is a considerable contribution from molecular diffusion along the column (longitudinal or axial diffusion) in the case of GLC, as diffusion in gases is many orders of magnitude faster than it is in liquids.

Compared with HPLC, the possibilities of varying the mobile phase are much reduced (to the extent of being virtually non-existent). However, there are a wide range of stationary phases available and temperature control is used as a powerful experimental parameter. Part of the success of GLC has been due to the development of a number of very sensitive detectors. Even the most straightforward of these, the thermal conductivity detector (which monitors the thermal conductivity of the column effluent compared with the pure carrier gas), can detect down to 10^{-7} g of sample with an upper useful limit of 10^{-2} g. The flame ionisation detector is the most commonly used detector and is shown in Figure 6.12(a). It operates over the range 10^{-9}—10^{-2} g of material and responds to almost all organic compounds. An alternative type of detector is the selective electron capture detector, which responds with great sensitivity to molecules

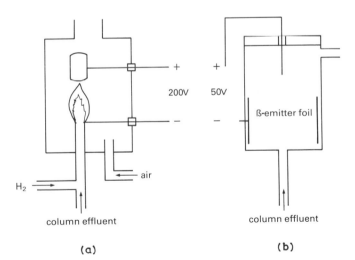

Figure 6.12 *GLC detectors.* (a) *The flame ionisation detector: burning an organic compound in the flame increases the ion population and hence increases the conductance.* (b) *The electron capture detector: electrons emitted from the ^{63}Ni foil ionise the nitrogen to give a background current, which decreases in the presence of electronegative molecules.*

containing halogens (down to 10^{-13} g) or metals. This detector is very useful for the analysis of pesticide and herbicide residues and organometallic compounds.

Qualitative and quantitative analyses are performed in much the same way as described earlier for HPLC. This includes quantifying the response of the detector to the particular component being sought. Peak area is the usual parameter taken for the basis of quantitative analysis.

One of the most common analyses by GLC is the determination of ethanol in human blood in connection with alleged drink–drive offences. However, an example illustrating the multi-component ability of GLC is shown in Figure 6.13. This is the

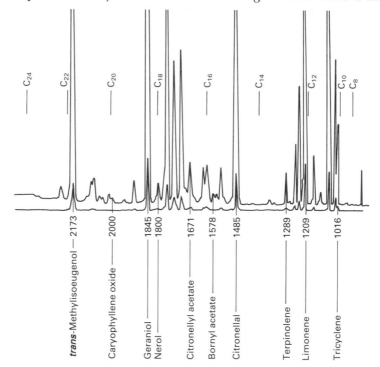

Figure 6.13 *GLC analysis of essential oil of citronella (Sri Lanka). A polyethylene glycol stationary phase was used with temperature programming from 75 to 225 °C at 2 °C min^{-1} and a flame ionisation detector. The two traces were obtained with different detector sensitivities.*

analysis of an essential oil (citronella, from Sri Lanka) used in inexpensive perfumery applications for household products. The chromatogram was obtained by temperature programming from 75 °C to 225 °C at 2 °C min^{-1}; the carrier flow rate was 30 cm^3 min^{-1}. The numbers under the peaks are the relative retention indices calculated relative to the C_{24} peak, assuming it has a value of 2400.

Recent Developments in GLC. As with HPLC, there has been a continued interest in developing the techniques and methods of GLC. Both types of chromatography are being made more powerful by the development of capillary columns. These are commercially available for GLC and have internal diameters of 0.1 to 0.2 mm and lengths of up to 100 m. The stationary phase is coated directly onto the wall of the tube. Very high efficiency numbers (up to 50 000) are possible, but the column capacities are very low and special techniques are needed to inject very small sample volumes onto the column.

Because of the large number of variables each instrument has, the reliability and ease of operation of both types of chromatograph have benefited from the development of microcomputer-controlled instruments. A computer-based integrator (to measure the areas under the peaks) makes quantitative evaluation a simple procedure as well. It is also possible to have automated sample injection devices and sample turntables that allow unattended overnight operation.

Probably the most powerful analytical technique of all in its ability to separate, identify, and quantify the components of a complex mixture is the combination of capillary column GLC and mass spectrometry (see page 50). As with the various spectroscopic developments mentioned earlier, a lot of research effort is being devoted to the development of chromatographic systems. There is no reason to suppose that the techniques have yet reached their full capabilities.

One group of separation techniques that is being increasingly used is based on the use of an electric field to move sample components.

ELECTROPHORESIS

Electrophoresis is a technique in which a potential gradient is produced along a gel bed (*e.g.* a 5 mm thick slab of starch).

Charged molecules move in the electric field (which may be quite high, with an applied potential difference of several hundred volts) to an extent governed by their electrical mobility. An initial compact sample zone thus separates into several zones if the components differ in their mobility. Human serum proteins may be separated in this way and screened for abnormal compositions characteristic of particular diseases. A more powerful (in the sense of analytical information) version of electrophoresis is isoelectric focusing. This is electrophoresis in a gel bed containing a pH gradient in which molecules will move until they have no net charge. For example, proteins will stop when they have reached the region which has the same pH as their isoelectric pH (the amino acids in the proteins have both acidic and basic functional groups – the isoelectric pH is that at which these are dissociated or protonated to an extent that the molecule has no net charge). Many more proteins in a mixture can be separated by isoelectric focusing than by conventional electrophoresis.

The basic technique may also be helped by using a gel which has a pore size gradient. A decreasing pore size restricts the movement of larger molecules, unlike gel filtration (page 140).

The separated components still have to be detected and measured. However, in some cases staining of the separated components with a coloured dye and then noting the presence or absence of a particular component or visually inspecting the relative amount of staining may be all that is required.

The success of electrophoresis methods depends on the availability of the appropriate media. These are now commercially available from a number of sources, and new developments are continually being reported by the manufacturers. There is considerable demand for these types of separation media as there are many analyses of a clinical or biological nature (such as 'genetic fingerprinting') which have such separations as an integral part. The continuing growth of the biotechnological industries is likely to result in increased usage of these methods.

FURTHER READING

'Ion Exchange Resins', 6th Edn., BDH Ltd., Poole, 1981.

P. Jenks and P. Wall, 'Thin Layer Chromatography', BDH Ltd., Poole.

A. Braithwaite and F.J. Smith, 'Chromatographic Methods', 4th Edn., Chapman and Hall, London, 1985.

Chapter 7

Tackling the Problems

In Chapter 1 I attempted to explain what sorts of problems analytical chemists are involved in solving, and in Chapters 2—6 I tried to explain some of the more widely used techniques that the analytical chemist makes use of. These chapters show that, in order to solve a particular problem, it is necessary for analytical chemists to have a good understanding of the chemical reactions and processes they use and of the physical principles on which the instruments operate. In addition, an understanding of the likely sources of error and of how these affect the overall uncertainty in the result is needed. So you might think that, apart from drawing attention to some analytical methods I have not mentioned so far (see Box 23) and pointing out that, like all scientists, analytical chemists need to be aware of the application of microprocessors and microcomputers in their subject area, this is the end of the story and you now know, as the title of this book puts it, 'what analytical chemists do'.

However, the constraints imposed by (a) the thermodynamics and kinetics of chemical and physical processes and (b) instrument performance are not the only ones under which analytical chemists operate.

WHAT'S THE PROBLEM?

Let's suppose that, as the manager of an analytical laboratory, you have unrestricted access to a wide range of instruments and methods: what information do you want about a particular analytical problem before deciding how to tackle it?

You should first find out what the overall object of the exercise

BOX 23 Analytical Techniques Not Mentioned

Name	Principle (measurement of)	Applications (determination of)
Nephelometry	Light scattered from sample	Suspended solids, $BaSO_4$, $AgCl$
Turbidimetry	Light transmitted through turbid sample	
Raman spectroscopy	Spectral distribution of scattered light	Molecular structure from information about vibrations
X-Ray absorption	As title	Heavy element in light matrix
Proton-induced X-ray emission (PIXE)	As title	Varied (thin solids)
Mössbauer spectrometry	Resonance fluorescence of γ-rays	Oxidation states and crystal structures (especially Fe)
Liquid scintillation counting	Light emission following electron collision excitation	β-Emitters (especially ^3H, ^{14}C, and ^{32}P)
Ion-scattering spectrometry (ISS)	Energy of ions scattered from collisions with surface ions	Elements in surfaces
Secondary-ion mass spectrometry (SIMS)	Mass spectrum of ions produced from surface on ion bombardment	Elements in surfaces and chemicals on surfaces
Refractometry	Refractive index	Concentration of single-component aqueous solutions (varied, but simple samples)
Interferometry		
Polarimetry	Change in direction of vibration of polarised light (optical rotation)	Sugars
Optical rotatory dispersion (ORD)	Wavelength dependence of optical rotation	Stereochemical features of optically active compounds

Circular dichroism (CD)	Wavelength dependence of extent of transformation of linear to elliptically polarised light	Stereochemical features of optically active compounds
Microscopy	Visual examination of magnified portions of sample	Useful first look at unknown materials. Fibres and paint
Thermo-gravimetry (TG)	Weight change as a function of temperature	Varied, polymer characteristics, composition of gravimetric precipitates, percentage purity, fingerprints of oils, waxes, fats, and explosives
Differential thermal analysis (DTA)	Temperature difference between sample and inert reference as function of temperature	
Differential scanning calorimetry (DSC)	Heat required to maintain zero temperature difference between sample and reference as a function of temperature	

is. Most likely, the sample you are being presented with is going to be either part of an on-going monitoring programme or a 'one-off' trouble-shooting analysis. In either case you will want to know what you're involved with as you may, for instance, be able to suggest a better way of achieving the desired objective. Secondly, you will want to know which determinands (*i.e.* the chemical entities being analysed for) are of interest and what the rest of the sample is likely to be. You should also know at what sort of concentration level the determinands and other (potentially interfering) species are likely to occur. You should find out what uncertainties in the results can be tolerated and what is going to be done with them. You will also want to know about the procedure for obtaining, storing, and transporting the samples before they arrive in your lab. Again, you may be able to suggest alternative procedures, but also you don't want to be held responsible for providing erroneous results when the samples are not representative of the material from which they were taken (because either the sampling procedure was inaccurate or the samples have undergone a chemical or physical change between sampling and analysis). You will want to know how quickly the results are required and how many, and how frequently, samples will be arriving. Depending on the financial framework you and your 'client' are operating within, it may be important to know how much the 'client' is prepared to pay for the analyses. Alternatively, the 'client' may want to know how much various options are going to cost.

The response to several aspects of the request concerning any particular analysis may well be that 'it can't be done' or 'it can be done, but...'. You, as the analytical chemist taking the job on, will have to be able to make these sorts of judgements and decisions right away, so that (hopefully) a realistic programme of analysis and reporting of results can be agreed.

CHOICE OF METHOD

Now that the problem has been properly stated, it is your job to devise an analytical method to solve it. An analytical method, don't forget, includes all the stages from recording the identification numbers as the samples arrive in the lab to putting your signature on a table of results. It may well be that your lab is responsible for taking the samples in the first place, and then you will have to give some thought to the problems of taking representative samples and of maintaining their integrity until they are analysed. Sampling is by no means a trivial problem – even seemingly innocuous homogeneous materials, like solutions, can be surprisingly inhomogeneous – and what about taking a sample from a train-load of iron ore to estimate the iron content?

Assuming that the new analytical problem is not just a minor variation on an existing problem that your laboratory is already handling quite satisfactorily on a routine basis, then you are going to do some research – but not the sort of research that means putting your safety glasses and lab coat on and embarking on an experimental programme. It may come to this in the end, but first you want to find out whether any other analytical chemist has had this problem or a similar one before. You may not be so fortunate as to be able to make a phone call to the person who has the answer to your problem (the cheapest and quickest way of solving problems – the trick is knowing whom to phone), and so you will have to turn to what has been written about this type of analytical problem.

Search the Literature

There is a large analytical chemistry literature in terms of books and journals, just as there is for other branches of chemistry and for other disciplines. There are also compilations of 'standard' methods (tested and evaluated by extensive collaborative trials) and other methods, used by particular sectors of the analytical community such as water, food, clinical, and metallurgical analysts. Maybe a solution to your problem can be found in one of these. You will need to know of their existence, of course, if they are not sitting on the shelves in your office or in your employer's library. Alternatively, you may need to search the primary literature, *i.e.* the large number of journals which publish the results of original research work in analytical chemistry. Some of this research is into the investigation of phenomena as the basis of new analytical techniques and some is into the solution of particular analytical problems. Some of these journals are general in their content (*e.g. The Analyst*) and some are specialised (*e.g. Journal of Chromatography*). Every now and again the journals publish reviews of particular analytical problems (*e.g. Analytical Methods for the Determination of Calcium and Magnesium in Wines*) or of particular analytical techniques (*e.g. Recent Developments in Detection Techniques for HPLC*), or even of both (*e.g. Standardised TLC Systems for the Identification of Drugs and Poisons*). So, just like any other scientist, you must know how to retrieve information from the relevant literature. It helps, of course, if you read the literature yourself on a regular basis and so have a reasonable idea of where things are to be found. If you work for a large company, there may be a specialised information department with all the latest forms of information technology to help you, provided that *you* can ask *them* the right questions.

As a result of your literature search you may find a suitable method or one that can be adapted. If not, you will certainly have some ideas about what you might try as the first candidate method. Now is the time to roll up your sleeves and put your lab coat on. Unless your analytical problem involves the determination of a minor or major component with the highest possible precision (and therefore requiring a gravimetric or titrimetric procedure to be used), your method will use an instrument as the final step.

Review Instrument Performance

In deciding which instrumental method you are going to use, you will have to know what the performances of the appropriate methods are. Factors to be considered include selectivity (*i.e.* freedom from interferences), multi-determinand capability, sensitivity, detection limit, speed, automated sample handling, ease of calibration, amount of sample, and pretreatment required. Alongside these factors the corresponding properties of the sample have to be considered: for example, how much is available, whether it will stand some chemical pretreatment, how likely it is that the pretreatment will contaminate the sample, whether a preconcentration step is needed. Having done all this, you should stand a reasonable chance of making a successful analysis. Any method being evaluated will have to be assessed for its precision, accuracy, and so on. Testing the accuracy can be a bit of a problem if you have no idea what the true value is for the concentration of determinand in the sample you are using for method development purposes. The way to approach this one, if you can't get a 'synthetic sample' of known concentration made up, is to analyse the sample, then add a known amount of the determinand and analyse it again. If you 'recover' 100% of the amount of the material added (plus or minus whatever the appropriate uncertainty is), you should have no reason to believe that the method is biased. This procedure is known as 'spiking'. It's not entirely foolproof, as the determinand already in the sample may be in a different chemical form from the 'spike' and thus have a different sensitivity by your method. You have to use your professional skill and judgement to decide how likely this is in the particular case you are dealing with.

The processes that go into the development of a method for a new problem are summarised in Figure 7.1, and a couple of examples are given in Box 24 of initial thoughts on what might be a suitable procedure (including financial constraints).

PROBLEMS WITH SOLUTIONS

Not the solution to the problem but a solution of the sample! You may have noticed how many of the commonly used analytical techniques require the sample to be presented to the instrument

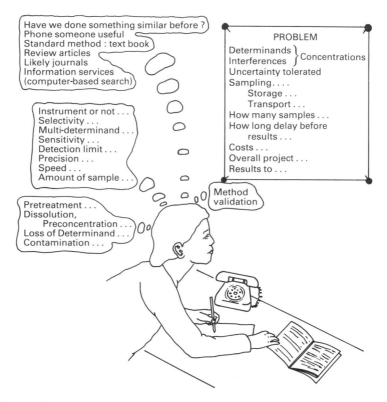

Figure 7.1 *The first stages of analytical problem solving.*

BOX 24   Initial Thoughts on Method Development

Suppose you are the chief analyst of a food-processing company which is about to produce a new product at several factories, namely a canned milk. The analytical problem concerns the amount of lead in the product, and it is your job to devise a suitable method for the determination of lead in canned milk. Your thought processes may go something along the following lines.

No problem with the sample; a suitable programme of taking the cans from the final production stage and after different periods of storage can be worked out. The statisticians will help with that. Trace metals nearly always involve an atomic spectrometry technique. We can't afford to invest in new instrumentation just for this analysis, so the method will have to be based on our existing trace-metal capability, *i.e.* flame atomic absorption spectrometry. However, the sample, although a liquid, can't simply be sprayed into the flame as the level of other sample components is too high to allow complete vaporisation during the residence time in the flame, and the likely level of lead will be below the detection limit of the technique anyway.

A sample preparation stage is therefore necessary to get rid of some of the sample components and produce a more sensitive response from the instrument.

So we could take quite a large amount of sample, say 25 g, dry it (probably overnight), ash the residue, dissolve what's left in nitric acid, and extract the lead as the complex with APDC into a suitable organic solvent, such as butyl acetate, which can be sprayed directly into the flame.

We'll need to take the standard solutions through the same procedure, but the calibration function should be reasonably linear, so we'll only need a blank and two standards. We will need to get the co-operation of our other three factories in order to get some kind of check on the validity of the results. I wonder if there's such a thing as a standard reference milk? Who would know that, I wonder?...

Now, suppose you are working in a Water Authority laboratory. Shortly, the lab will have to start analysing all samples for both nitrate and nitrite. Several hundred samples of all sorts of water (river, bore-hole, *etc.*), varying in quality from potable to grossly polluted, arrive in the lab every week. Your thought processes may go something like the following.

The large number of samples means that, if we are to keep pace with the requirement, an automated method must be used. At present there is no spare capacity on the existing automatic analysers, so some equipment will have to be purchased. We can't afford another 20-channel analyser, so it will have to be something relatively low cost, probably dedicated to this particular analysis. The method of analysis has got to be compatible with the automated format, and that means flow-through pretreatment and measurement. This means that a spectrophotometric method or possibly an electrochemical method will be used. Ion-selective electrodes are alright for nitrate, possibly, but what about nitrite? A spectrophotometric method will probably be better because the same chemistry can be used for both anions. So what is needed is a method for nitrite, as I am sure that nitrate can be reduced fairly easily to nitrite. Let's have a look at Williams' book. Okay, so no real problem about the chemistry; the Griess–Ilosvay method seems to be the one. Note that some versions use reagents that are now known to be carcinogenic. The segmented-flow method is going to be too expensive in terms of capital equipment. I wonder if it could be done by flow injection analysis? Sample preparation is no real problem, just filter out the particulate material that could block the tubes. Now where did I see that review on FIA methods?...

as a solution. Unfortunately, the particular samples you may have to deal with may be quite difficult to get into solution.

As with taking samples in the first place, this is not a trivial problem, and again the whole success of any subsequent chemical and measurement steps will depend on whether all of the determinand can be brought into solution. However, as solid samples are so common, analytical chemists have already given a lot of thought to the development of dissolution procedures, and these are readily available in standard text books. Some procedures are summarised in Box 25. Analytical chemists have to use their professional skill in deciding which dissolution method to adopt, how likely the procedure is to result in the loss of the determinand, or whether the reagents used are themselves contaminated with the determinand. Strong acids, for example, are particularly difficult to obtain free from traces of metals. Sometimes the specific chemistry of the required species has to be considered, *e.g.* solder (an alloy of lead and tin) may be dissolved in concentrated hydrochloric acid in the presence of edta (which prevents the precipitation of $PbCl_2$), and there are other situations in inorganic analyses where complexing ligands can be usefully used.

There are cases, of course, where the solid material is used directly. In the determination of the C, H, N, *etc.* content of an organic compound the solid sample is burned in a stream of oxygen and the amount of CO_2, H_2O, *etc.* produced is measured. The determination of gases in metals is another example where the solid sample is decomposed in the instrument as part of the analysis. X-Ray fluorescence and arc/spark atomic emission spectrometries both require a solid sample (about the elemental composition of which they provide information).

FINANCIAL CONSTRAINTS

So far we have been assuming that there is an unlimited amount of analytical instrumentation available to help solve the various problems coming to the analytical laboratory. However, in any commercial organisation the analytical support for the overall operation will be operating on a budget, just as other parts of the business will be. It will be up to the analytical chemists in the organisation to explain what is needed in the way of staff, equipment, *etc.* to their senior management, who may not even be

BOX 25   Getting the Sample into Solution

The kinetics of any dissolution procedure are improved if the sample can present a large surface area to the solvent. If the sample does not have an open, porous structure or is not a finely divided powder, then the first step may be to grind, chop, mince, macerate, *etc.* (grinding under liquid nitrogen can be a useful technique for 'organic' samples).

It may not be necessary to get all the sample into solution, *i.e.* the dissolution step could be used as a separation step. On the other hand, if it's all gone into solution, then the determinands can't be left as a solid residue. They may, of course, have gone up the chimney!

ORGANIC COMPOUNDS

Generally speaking, these won't stand very severe dissolution procedures, particularly those involving heat and oxidising conditions. Dilute acids, alkalis, and organic solvents are probably more suitable than hot, concentrated acids. It is possible to set up a continuous solvent extraction procedure whereby the sample is extracted with refluxed solvent so that gradually all soluble components are transferred to the still pot.

Unwanted organic components of a sample can be removed by burning in oxygen, heating in a furnace (dry ashing), or oxidising with hydrogen peroxide or hot nitric or perchloric acids (wet ashing); considerable care is needed with the latter species. Don't forget that the analyte species may be relatively volatile, even if an 'inorganic' component is being looked for.

INORGANIC COMPOUNDS

These usually require acids; the order of severity is approximately HCl, H_2SO_4, HNO_3, HCl–HNO_3 (3:1 gives 'aqua regia'), $HClO_4$ (combined with others including H_3PO_4), HF (nasty). Platinum dishes *etc.* may have to be used. Anything that won't go in these acids will need to be attacked by a molten salt (called a 'flux' or 'fusion mixture'), NaOH, $LiBO_2$, LiF–H_3BO_3, or Na_2CO_3. Sometimes high pressure is needed, so samples and solvent are placed in a gas-tight, teflon-lined, stainless-steel vessel with a bolt-on lid and heated. This device is known as a digestion bomb! After dissolution in a fusion mixture the solidified melt is dissolved in a suitable acid.

chemists. It's important therefore that analytical chemists can write a clear, concise, understandable, convincing case for the equipment needed and are able to defend this when it is queried.

Sorting out the cost-effectiveness of analytical equipment can be quite tricky. It's not just a simple matter of the capital cost of the instrument, even though it may be the only way to provide useful answers to the particular problem. Large, expensive instruments can usually be 'hired' from another industrial firm or from a university or polytechnic (the samples have to be taken along to where the instrument is located). This means that just for the occasional sample you don't have to buy the instrument. However, as the number of samples increases there will come a point when it may be better to buy. It's also possible to buy other analytical expertise, and there are several companies around the country that will carry out a whole range of analytical work on a contract basis. The analytical chemists who run such companies are known as consultants, and their analytical skills and professional integrity must be of the highest order. They face the challenge of having their livelihood depend on their skill in solving analytical problems.

In addition to capital cost there will be a number of other costs associated with using an instrument, such as running, servicing, and repair costs. For example, does the instrument require a special air-conditioned laboratory and a specialist technician to run it? Offset against this is the number of samples or the number of analyses on each sample that the new technique can deal with, thereby relieving pressures on existing staff, increasing the analytical lab's efficiency, decreasing sample turn-around time, and so on.

Analytical instruments are continually being improved, and new types of instruments appear from time to time as manufacturers turn the research analytical chemists' discoveries into working instruments. Eventually, decisions have to be taken about when to replace an instrument with an updated model or when to invest in a new technique.

KEEPING UP-TO-DATE

Analytical chemists, just like other scientists, have to keep themselves up-to-date with developments in their subject that have taken place since the finish of their formal education. Scientific and technological developments are moving at such a pace these days that during a scientist's working lifetime of about forty years there are bound to be many major advances. Many of the instrumental techniques described in this book that are now in common use have been developed over the last forty years. For analytical chemists now nearing the end of their working careers all the major instrumental methods have been introduced since they completed their formal education. There is no reason to suppose that the next forty years will not see developments just as fundamental, and so analytical chemists have to be prepared to go on learning long after their formal education has ended. One of the ways in which they can do this is to join a professional society (see Box 26).

BOX 26   The Royal Society of Chemistry

Keeping scientists abreast of developments in their subject area is one of the major functions of the large number of professional societies. The chemists' needs are catered for by The Royal Society of Chemistry and the particular needs of analytical chemists by its Analytical Division. The Analytical Division organises a full programme of meetings, symposia, and conferences on the whole range of analytical topics, at locations all over the country, to inform and educate chemists about recent developments and future trends, *etc.* These meetings also provide analytical chemists with the opportunity to meet each other and discuss problems (and build up their list of contacts when it comes to deciding whom to phone!). The Royal Society of Chemistry also publishes specialist journals on analytical topics as well as a general-interest journal called *Chemistry in Britain.*

In addition, some sections of the analytical community have their own specialist professional society such as The Forensic Science Society, The Association of Clinical Biochemists, and The Association of Public Analysts. Only The Royal Society of Chemistry can award the professional status of 'Chartered Chemist', a designation that trainee analytical chemists should work towards if they wish to be considered professional analytical chemists. Further information can be obtained by writing to: The Membership Manager, The Royal Society of Chemistry, Burlington House, London W1V 0BN.

WHAT TO DO WITH LOTS OF SIMILAR ANALYSES

Perhaps one of technology's most fundamental and far-reaching developments in recent years has been that of automation. Virtually no aspect of a domestic or industrial process has been exempt from such developments as our society strives to make lives more comfortable and industry more profitable. Although analytical laboratories don't yet look like automotive production lines, there are a number of small, bench-top, versatile robot arms on the market, capable of performing some of the more routine operations such as weighing, stirring, shaking, *etc.* However, developments at present have been along the lines of (*a*) automated sample preparation and presentation to instruments, (*b*) increased ability of instruments to select operating conditions and process results correctly, and (*c*) increased use of microcomputers to collect data from instruments, perform calculations, and print reports.

Some instruments have the ability to introduce samples automatically and measure them from a special 'turntable' loaded up with samples, and thus can be left running over lunch-breaks, overnight, or even over the weekend. Proper consideration has to be given to keeping the instrument calibrated, and calibration standards will need to be interspersed with the samples.

A number of instruments will handle some of the sample preparation as well. If the chemistry to be performed consists only of adding reagents, then once the sample is in solution the instrument takes the sample on a 'conveyor belt' around various work stations where reagents are added, heat is applied, and so on. Finally, the appropriate measurement, most likely spectrophotometric, is made. These types of analysers are known as 'discrete' analysers because they only analyse for one determinand in the sample and the samples in solution are transported around the system as separate entities.

Another type of automatic analyser is based on the principles of continuous flow. In these systems the sample is transported in a narrow-bore pipe and reagents are added from other pipes which join the main one. Usually these instruments are constructed on a modular basis, so it is possible to perform a variety of different assays with the same basic building blocks. Also, a greater variety of manipulations can be automated, such as dialysis, gas diffusion (converting the analyte into vapour and separating it by diffusion through a membrane), liquid–liquid extraction, and ion exchange (packed-bed reactor). In one version of the continuous-flow principle, the samples are separated by air bubbles. Within the last ten years it has been shown that, if smaller tubes and sample volumes and lower flow rates are used, the bubbles are not necessary. Examples of the two types of system are shown in Figure 7.2. Multi-channel flow analysers are commercially available and can handle up to ten different determinands. Over two hundred samples per hour can be analysed, so the number of

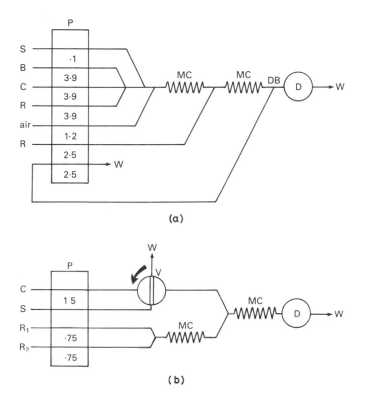

(a)

(b)

Figure 7.2 *Continuous-flow analysis.* (a) *Air-segmented continuous-flow analyser for magnesium: P, pump (flow rate in each line given in cm³ min⁻¹); S, sample from turntable; B, buffer; C, carrier stream (0.1M KCl); R, reagent (o,o'-dihydroxyazobenzene); MC, mixing coils; DB, debubbling point; W, waste; D, flow-through detector (fluorimeter).* (b) *Flow injection determination of phosphate: V, injection valve (200 mm³); R₁, molybdate reagent; R₂, ascorbic acid reagent; other symbols as for (a); D, spectrophotometer.*

determinations possible is getting on for 2000 per hour.

Continuous-flow analysers provide analytical chemists with the challenge of devising (a) chemistries that can be adapted to the flowing streams and (b) flowing systems to cope with particular chemistries (procedures such as evaporating to dryness or heating at elevated pressures do not lend themselves to flowing systems).

Some industrial processes operate more efficiently if the concentration of certain components can be continuously monitored 'on-line' (*i.e.* at the location of the reaction in question). This means that analytical equipment has to be built into the reaction vessels, transfer lines, *etc*. The output from such instruments may be used to control the process directly, raise alarms if the process is going out of control, and so on. Analytical chemists may therefore be involved at the design stage of an industrial plant to advise on what analyses are possible, what the best location for sensors may be, what sort of equipment to install, and what to do with the information they provide.

Continuous on-line analyses will also be used to monitor other aspects of a chemical manufacturing process such as the level of potentially dangerous substances in the work-place atmosphere and the concentration of potential pollutants in the various effluents.

FURTHER READING

R. Smith and G.V. James, 'The Sampling of Bulk Materials', The Royal Society of Chemistry, London, 1981.

D.T.E. Hunt and A.L. Wilson, 'The Chemical Analysis of Water', 2nd Edn., The Royal Society of Chemistry, London, 1986.

J.C. Van Loon, 'Selected Methods of Trace Metal Analysis', Wiley, New York, 1985.

W.J. Williams, 'Handbook of Anion Determination', Butterworths, London, 1979.

D.J. Huskins, 'Quality Measuring Instruments in On-Line Process Analysis', Ellis Horwood, Chichester, 1982.

Subject Index